AF722877

Digital Learning Series
A Series by Post-Technology Educators

To the extent that education is the engagement of individual learners to knowledge, the Internet has changed everything because knowledge has moved to a new location. What is known by humankind is now located primarily on the Internet, and what is not there today will be soon.

The great migration of knowledge into cyberspace has kicked off a technological revolution in education. Teachers have marched off to computer classes. Librarians have been retrotrained as technology specialists. The "crayon set" can type before they're in kindergarten, and coding is a second language for the generation of kids now in high school. Millions of dollars have been spent wiring schools, and millions more are being spent reworking the wiring to broadcast the Internet to mobile receivers.

All the technological advances, both in equipment and user skills, have certainly been crucial to engaging knowledge on the Internet. But there is a further necessary component to the new education era introduced by the Internet. This book series swings the focus from the technology of connectivity to the digital form and substance of the knowledge engaged by learners on the Internet. Where is the information and how do we find it? In what form does it appear? How do we learn it? How do we teach it? These are real questions encountered every day in classroom teaching, independent study, and the preparation of online learning materials.

We can answer these questions by realizing that the interesting challenges today are no longer technical. True innovation is concerned with creating learning resources in the digital milieu. It is technically possible to link to almost anything on the Internet, but how do we choose what to link to in order to create an effective learning environment? Teaching and learning knowledge in its new location require tools that are *cognitively* new, though no longer *technically* so. How does interactive multimedia connected to a stunningly rich context web make learning grammar or geography or genetics different from studying these subjects in an isolated classroom or book? How can I implement the advantages and minimize the disadvantages in my own teaching and learning? These and many other such immediate and important questions arise once we become competent on the technical side.

The authors who have written books for this series are active participants in the challenges of engaging individual learners to the Internet, describing the reconnection to knowledge in our post-technology education world.

Judy Breck, Series Editor
New York City
September 2003

Digital Learning Series
Edited by Judy Breck

1. Jennifer Cordi. *Engaging Knowledge: The Inference of Internet Content Development and Its Meaning for Scientific Learning and Research*. 2004.
2. Noah Kravitz. *Teaching and Learning with Technology: Learning Where to Look*. 2004.

ENGAGING KNOWLEDGE

The Inference of Internet Content Development and Its Meaning for Scientific Learning and Research

Jennifer Cordi

Digital Learning Series, No. 1

ScarecrowEducation
Lanham, Maryland • Toronto • Oxford
2004

Published in the United States of America
by ScarecrowEducation
An imprint of The Rowman & Littlefield Publishing Group, Inc.
4501 Forbes Boulevard, Suite 200, Lanham, Maryland 20706
www.scarecroweducation.com

PO Box 317
Oxford
OX2 9RU, UK

Copyright © 2004 by Jennifer Cordi

All rights reserved. No part of this publication may be reproduced, stored in a retrieval system, or transmitted in any form or by any means, electronic, mechanical, photocopying, recording, or otherwise, without the prior permission of the publisher.

British Library Cataloguing in Publication Information Available

Library of Congress Cataloging-in-Publication Data

Cordi, Jennifer, 1971–
Engaging knowledge : the inference of internet content development and its meaning for scientific learning and research / Jennifer Cordi.
p. cm.— (Digital learning ; no. 1)
Includes bibliographical references and index.
ISBN 1-57886-088-1 (pbk. : alk. paper)
1. Science—Study and teaching. 2. Internet in education.
3. Problem-based learning. I. Title. II. Series.
Q181 .C628 2004
507′.1—dc22 2003020473

∞ ™ The paper used in this publication meets the minimum requirements of American National Standard for Information Sciences—Permanence of Paper for Printed Library Materials, ANSI/NISO Z39.48-1992. Manufactured in the United States of America.

to Dan and Jessica—

Any journey is made sweeter
with the love and support of siblings.

CONTENTS

PREFACE

Unlike any other human construct, the knowledge we share has unlimited potential for growth and development. Our continued quests for answering age-old questions are culminating in a generation of new inquiries and aspirations for the future. But, seemingly, as our knowledge grows, fields of study are becoming more and more fragmented. We focus on distinct components or levels of complex systems, trying to understand their characteristic properties and functions. For instance, these days, there is not a single discipline of "biology," the study of life. Instead, there are the studies of microbiology, immunology, genetics, molecular biology, physiology, anatomy, biochemistry, histology, ecology, sociobiology, pathology, paleontology, and so on. These subfields, in turn, are comprised of further specialized areas of study, all of which have emerged from our efforts to make scientific knowledge more tractable and comprehensive.

However, despite many of our ambitious attempts to scientifically pursue the true nature of life, I often think that we are simply manipulating separate pieces of a gigantic jigsaw puzzle, focusing on these pieces rather than on the big picture. To our benefit, a macroscopic view of the puzzle's picture might render properties of life that can be elucidated only by understanding the cumulative interactions of the pieces—by knowing how the pieces fit together rather than knowing the pieces themselves.

So much scientific knowledge has and continues to be accumulated that it demands the coalition of abundant disciplines and forms of literacy so that we can continue to make progress and sense of it all. Being literate today demands not only one's accurate knowledge and use of language but also a basic understanding of science, mathematics, and technology. Moreover, literacy today requires an understanding of the structure and function of computers and the Internet. This latter form of literacy is now commonly referred to as "digital literacy," and it is rapidly becoming one of the most important skills for the advancement of learning and research.

This book is intended to showcase the significance of digital literacy as it pertains to the explosive growth and expansion of Internet content. Why should we care that information is continually being developed and disseminated all over the world via the Internet? I have talked to so many people over the years who pose this same question in slightly different ways: "What's the big deal with the Internet; isn't it mostly filled with junk?" "Shouldn't we be trying to control the Internet rather than let it grow?" "Technological advancements come and go, and the Internet is more aggravating than anything else, right?" These questions are important, and they need to be addressed with accurate explanations of what the Internet is, how it was created, and how it can be used as a powerfully constructive communicative and research-oriented tool.

As I hope to convey in this book, what makes the development of Internet content so significant, influential, and relevant to understanding both our history and our future is its unmatched potential to bring people, knowledge, and information together. Everything we have come to know throughout human history is steadily finding a place of permanent residence and unhindered accessibility on the Internet. The implications of this technological advancement for human learning, research, communication, and collaboration are incredibly exciting.

This work is meant for students, educators, researchers, and anyone interested in lifelong learning—learning that extends far beyond the confines of the traditional classrooms or course syllabuses and actively progresses throughout our entire lives. I hope to offer a new understanding of the structure and function of Internet content and how it can be accessed and used to augment learning and research methods. This book is not about techniques of web development, it is not a showcase of the best scientific and educational websites on the Internet, and it is not a compilation of online lessons or activities for integration with conventional curriculum models. Rather, it is intended to help us understand how the information age is influencing our basic thoughts, needs, and practices concerning education, scientific research, knowledge acquisition, and the growth of human intellect. I hope to show how Internet content can provide a new means for solving old problems that have long plagued scientific education and research while at the same time opening doors for new advancements and opportunities. In essence, this book is about the future of learning.

We are constantly hearing all that is bad about the Internet. The time has come to explore what is good. As people from all around the globe begin and continue to access the Internet, exchange their information, and share their ideas, inevitably much of what constitutes our world will be accessible online. Although the worst and most devious of human activities may be showcased on the Internet, it is also true that the most amazing of our accomplishments, hopes, and dreams exist within the realm of cyberspace. We cannot ignore the good for fear of encountering the bad.

ACKNOWLEDGMENTS

In parts of this book, I reflect on a few past experiences that I have found to be extremely influential and relevant to my ideas about online learning. I do not mention the identities of people involved but rather focus on the particular events and personal interactions that have come to represent positive forces in my life. Here, I would like to acknowledge those people who have truly helped me mold the long-accumulated vision that I have brought to this work as both an educator and a scientist.

I am forever grateful to Judy Breck for giving me the opportunity to pursue this work. Her guidance, wisdom, and camaraderie have been solid foundations from which to launch my most recent explorations of lifelong learning. Our discussions about the future of Internet content development have been some of the most motivating, enthusiastic, encouraging, and inspirational for me as I seek to advance my own potential and contributions to the unending stream of human knowledge. I am grateful to have had the opportunity to work with her, and she has become a lasting friend and colleague.

Many of the perspectives and values presented in this book I developed as a young graduate student. My progress as a student was enhanced and nurtured by the wonderful relationships I had with members of my dissertation research committee. They took me and my work seriously, treated me as an "equal" within the scientific community, and, constantly yet constructively and respectfully, challenged my views and methods. My research adviser, Bill Stein, never stopped encouraging me from pursuing my own research interests, asking my own questions, and finding my own answers. He helped me gain skills needed for critical thinking, writing, and analysis. He also helped me understand the importance of conceptual, analytical, as well as historical views to scientific inquiry.

My botany professor, Pat Bonamo, once told me that I was a born teacher. When she said this, I was overwhelmed with delight because I always viewed her as the greatest role model that any teacher could ever have. Throughout graduate school, she encouraged my interests in teaching and helped me to understand that working directly with students was a critical asset to any scientist's teaching as well as research agendas. Students help keep you honest, on the ball, and forever asking the question, "Why?"

I am grateful to have had the opportunity to work with such adventurous, visionary, and knowledgeable faculty members of the science department at Bard High School Early College. When I came in as a rookie teacher last year, they provided tremendous guidance and support. We worked together to tackle so many challenges, and I learned a great deal from them. In particular, they helped me to see different points of view about education and to try new techniques and approaches to learning.

Last but not least, I am grateful for my time spent with students at Bard High School Early College. In many ways, I learn more from them than anyone else. Each day is a new adventure in getting to know them and how they learn best. I will always remember one student in particular who started off the academic year with an angry, hostile, and apathetic attitude about school and learning in general. Her attitude persisted for much of the year. But, as I continued to actively care about her progress, she eventually developed into a serious, conscientious, and respectful student. When I reached out to her and she reached back, it was one of the most rewarding and inspirational experiences that I have ever had.

PART I

NETWORKS OF THOUGHT

In an extreme view, the world can be seen as only connections, nothing else.

—Tim Berners-Lee, inventor of the World Wide Web, 1991

1

PIONEERS OF A NEW FRONTIER

Emerging technologies are forever changing the way in which we obtain, process, and transfer information. Portable devices such as laptop computers, personal digital assistants (PDAs), and cell phones with wireless Internet connections are providing us with instant access to, literally, a world of knowledge. Remarkably, every form of information that has previously existed on printed paper or through audio or visual transmissions now also exists or is capable of existing in electronic form on the Internet. And, in an unprecedented fashion, electronic information on the Internet transcends far beyond the distinct, conventional formats of media exchange, showcasing multifaceted forms of live discussions, text messaging, streaming video, animations, data displays, calculations, and simulations. Never before have we been able to communicate so much to so many in so little time. Now and for the first time in our history, people from around the globe are obtaining equal access to information and technology.

The influx and spread of Internet content is immense and surpasses that of any other form of media, past or present. Since the origins of written language, the historical trend toward global information exchange continues to influence the way in which we communicate and articulate our ideas through social and political interactions, artistic expressions, and scientific inquiry. However, the course of human expression and dialogue is now taking a sweeping new turn. We are truly living in a fascinating and pivotal time in human history, and we are only beginning to understand the implications of online information access to improving our opportunities for better education, domestic and foreign relations, economic developments, and technological advancements (Breck, 2004).

I first became aware of the power of Internet content while working as a content manager for a leading online educational company in the summer of 2000, bigchalk.com. Among the many subscription-based, online learning resources (such as lesson plans, assessment tests, and periodical databases) the company made available to students, parents, teachers, and librarians, it also offered a free database filled with the best educational websites on the Internet. This database was showcased and maintained by the developers of a subsidiary site of bigchalk called homeworkcentral.com. Students could log on to homeworkcentral and browse dozens of academic subject areas. I managed the science portion of the homeworkcentral database, constantly surfing the Internet and looking for the best-quality scientific websites in cyberspace: sites not only that had the latest and most reliable scientific news and information but also that went far beyond the presentational formats students would regularly encounter in a textbook or journal article.

While at bigchalk, I worked with other scholars who were in charge of aggregating content in other subject areas to organize the links within the database in a way that would encourage students to browse online content in an interdisciplinary manner. For example, if a student began a query about Albert Einstein, he would begin an online journey that might take him not only to various sites about relativity, light, space, the periodic table of elements, and the most basic to most advanced principles of modern physics but also to sites about Einstein the man and his times, the history of Europe during the early twentieth century, World War II, and the Allied challenges, strategies, and conflicts. This was my dream job at the time; exploring the Internet, being awed by its immensity and its power and potential for connecting ideas. I was absolutely captivated with the mission of sharing my newly discovered online gems with students of all ages, needs, and abilities who equally wanted to explore and learn from the Internet.

Several months after the stock market crash of 2000, the company downsized, and I was let go with academic content managers, editors, members of programming staff, and others. I was devastated, to say the least. Not only did I lose a job that I absolutely loved, but I became obsessively concerned about the student users of the database. I kept asking myself, "Who will showcase the wonders of the Internet for them? How will they find the incredible websites that are out there?

How will they find the particular sites that are best for them? How will they come to know the power of the Internet, and how it can maximize their individual learning potentials?" Although my mind was deeply preoccupied with these questions for some time, I finally realized that no one can showcase the Internet and all its wonders for students with the hopes of their fully grasping all that is has to offer their learning needs. Instead, students have to discover it all for themselves. The personal rewards and global contributions of self-paced, self-directed, and self-motivated academic explorations through cyberspace is what this book is about.

Throughout my time spent as both a student and a teacher, I have come to understand and experience firsthand the educational endeavors of online learning. This has happened not by my attempts at becoming "tech savvy," building websites, or learning countless computer programming languages (to this day I have yet to even learn html) but rather by being an active user of Internet content, overcoming a "fear of computers," and putting aside educational biases and looking to see what the Internet has to offer learning that no other resource, instructor, or institution has had or ever will have.

During my graduate school years in the late 1990s, I think that I was probably one of the last of my generation to still exhibit computer phobia. It took semesters for me to become acquainted with regularly using the basic programs of Microsoft Office, let alone using the Internet for any meaningful purpose. Luckily, when the time came such that I could no longer ignore the benefits and needs of software use, I was a quick study in the realm of computer competency. (A scientist can communicate little to nothing today without having digital literacy.) Now, I cannot remember or even imagine life without computers and the Internet. Whenever I see an old movie, an old piece of television footage, or an old picture from a scrapbook or newspaper, the first thing that pops into my head is, "Wow, life before the Internet." And I cannot help wondering, "What must it have been like when the world's information wasn't at your fingertips?"

Before I begin discussing at length the new age of online information exchange and all its sensational renderings, let me first define what I mean by "information." In my view, information is comprised of any type of discrete data (such as statistics or measurements) as well as any written, audible, or visual form of communication. "Knowledge," by contrast, comes from our interpretation of the information, by making associations with other relevant pieces of data and collaborating and sharing our ideas with others. This relationship between information and knowledge is based on a "constructivist's" point of view about how we come to understand ideas and events and thus acquire knowledge.

Philosopher Richard Rorty (1991) has suggested that knowledge is created in the minds of individual learners, within the range of personal experiences and interests. Although constructivism as an applied method of learning is not new, I believe that constructivist theory and practice are beholding new relevance and application with the dawn of Internet content development and online information exchange.

More and more people around the world are accessing the Internet not only for the immediate retrieval of information in the form of stock quotes, road maps, or recipes but also as a primary means for learning, discussing, exploring, and sharing what they know. The abundance of information that resides on the Internet is becoming available to all persons, regardless of status or geographic location (Breck, 2001, 2002, 2003a). As a result, we are perceiving our world as a much smaller place with permeable boundaries. This permeability is sustained by a prevailing conduit of instantaneous information exchange. The prevailing conduit is the Internet, and through its accelerated and expansive use, the ways in which we research, communicate, and learn are just beginning to acquire new meaning and relevance.

The ongoing process of exchanging information and acquiring knowledge over the Internet is what I refer to as the "development of Internet content." Advancements in wireless technologies, coupled with a recent explosion of Internet content development, have sparked innovative approaches to research and learning as seen in some of the fresh, new efforts toward reevaluating teaching techniques, developing semantic search engines and synchronous, interactive tools of virtual schools and universities. Although many people are clearly acknowledged as leading developers of Internet content (for example, Breck, 2001, 2002, 2004; Berners-Lee, 2000; Berners-Lee, Hendler, & Lassila, 2002), we must realize that all of us are pioneers of this vast, new frontier.

The Internet represents an endless universe of knowledge, explored by individual users connecting various sectors or points of interest. We may think of these points as the individual websites or pages, whereas sectors are clusters of associated points making up what I refer to as online "information modules." These modules are distinguished by the unique semantic pathways of individual users. Precise pathways are created by the sequential "clicks" a user makes between each point of his Internet exploration. Information modules are user specific and exist only in relation to a particular user's semantic pathways. These modules are transient and dynamic. They are constructed or degraded depending on a user's exploration objective. Each time we use the Internet to discover facts, research, or share knowledge, we build unique information modules and contribute to the development and overall significance of Internet content. As a result of its continual exploration, the Internet is forever expanding and evolving; new points are added or removed, and new sectors are generated. We are becoming a global community of learners, of information seekers and sharers, as we explore the Internet and engage its means for augmenting human knowledge.

Students of science have much to gain from participating in this boundless exploration of Internet content. The Internet's expansive (forever growing) and evolvable (forever changing) nature is a vital aspect of discovery learning and scientific methodology. Such modes of inquiry are not new to educators and scientists. Rather, for many they represent the proper and most effective ways of gaining knowledge and providing explanations of natural phenomena. Scientific methods, whether inductive or deductive, involve the testing of hypotheses, the collection and analysis of data, the communication of results, and the formulation of theories. Discovery, or "inquiry-based" learning, relies on a student's critical thinking skills in decision making, problem solving, evaluating, and comparing data.

Students work at different paces and have different cognitive strengths, abilities, and academic interests. They may view problems differently and provide different ways of expressing the same solution. Through self-paced and self-directed modes of inquiry, students of all ages may demonstrate their own desire to ponder and learn, to design experiments to answer their own questions about natural phenomena, and to use mathematical expressions or computer programs to model real-world systems and processes. Never before has there been a medium more capable of enhancing the applied outcomes of discovery learning and scientific inquiry than the Internet. Although these methods of learning may apply to all academic disciplines, I will focus on their powerful influence in science. This is due only to my own personal bias. I am a biologist by formal training, and the similarities between discovery learning and scientific methodologies are most apparent to me. However, this is not to say that similar approaches cannot be followed in other disciplines.

Internet content is endless, dynamic, interdisciplinary, and easily accessible. The way in which we use the Internet to investigate and gain scientific knowledge is the focus of this book. But as you will see, the critical thinking and problem-solving skills that we may gain from our learning endeavors come not from a specific type of content but rather from the actual process of engaging knowledge via the Internet.

Taking advantage of all that the Internet has to offer requires an understanding of the structure of the World Wide Web and how information may be retrieved and transmitted in a resourceful, responsible manner. Artificially intelligent search engines are beginning to have an impact on tasks such as these. In chapter 2, we will become familiar with the development of artificially intelligent search engines and the three-dimensional structure of web networks and their inherent "lack of center" and nonhierarchical frameworks for information storage and retrieval. As we will see, the structure of web networks is reminiscent of our own semantic neural networks. Moreover, clicking through points of hypertext and forming user-specific information modules is intriguingly similar to how our minds think and how they process information.

In chapter 2, we will also see how the Internet is a unique research and learning tool characterized by dynamic, context-based hypertext. Progress in science relies on the synthesis of new, collaborative, and up-to-date information. As such, scientific resources have come to rely more and more on a fast and dynamic medium for research and communication purposes. We will see in this chapter the unique capacity of the Internet for sharing knowledge, creating context, and enhancing relevance for scientific studies.

At this point, it is important that I make clear that the focus of chapter 2, as well as this entire book, is not about the intricacies of web technology or a review of educational websites, scientific websites, online references, or online learning activities. There is already a multitude of quality publications on these topics, a few of which I make reference to throughout this book. Rather, my aim is to express a way of "thinking about" and "doing" science based on the innovative tools offered exclusively by the Internet. Chapter 3, for example, discusses how different philosophical views and overall objectives of science can be influenced by the everyday use of Internet technology.

Never before has the need to move beyond "reductionist thinking" in science toward "context-based" inquiry been more apparent to me than with the influx of Internet content. The reductionist approach to science aims to discover and explain the structure and function of the individual components, subparts, or "building blocks" of a system. In contrast, what I consider a context-based approach deals not only with the subparts of a system but, more important, also with the relationships between the parts. My major concern with reductionist science is that by taking details out of relevant context and removing them from direct or even indirect associations, significant information is lost and explanations may remain problematic. In chapter 3, I focus on the use of context-based hypertext and how it may ultimately change the way in which we study, hypothesize, and conduct research. We will begin to view problems from multiple vantage points, ask bigger questions, and seek comprehensive and universal explanations. By formulating online information modules through user-specific semantic pathways, we may come to better understand concepts of collective behavior and emergence: what parts of a system do together, by virtue of their relationships, that they would not do by themselves.

As obvious to most people, the Internet is a medium with far-reaching capabilities. By design, Internet content is global and interdisciplinary. Knowledge and information are truly shared instantaneously around the globe. Collaborations may occur among scientists across continents and across various levels and disciplines of research. In chapter 4, I discuss how access to direct sources, hands-on simulations, and context-based hypertext augments scientific learning, research, and instruction by transcending standards of core curricula. No longer will science educational practices be "tracked" within schools (such as grouping students in separate classes according to their intellectual strengths or ability levels), be viewed as strictly "American" or "European," or have any specific political or social relevance. Rather, we will begin to have a new "universal" perspective on the meaning of education and learning.

Much of the grand, global influx of scientific information onto the Internet has occurred through a process of "repositioning": placing previously published material or software onto an electronic web page that is made to resemble the original, communicative, and functional format of the publications. Chapter 5 focuses on examples and uses of content repositioning, such as e-books, periodicals, virtual classrooms, calendars, and library catalogs. Surprisingly, in addition to information itself, some scholars have also sought to reposition full-blown methodologies of teaching and research onto the practices of Internet content development and use. For example, aligning Internet content to state education standards and building two-dimensional hierarchical search engines and classification schemes are popular online practices today. However, in my opinion, they are not at all compatible with the structure and function of the Internet.

I find that developers who seek only to reposition published information or re-showcase methods of teaching and research are completely missing the point when it comes to utilizing the Internet as a media outlet or learning tool. In chapter 5, we will see how the Internet, through its dynamic makeup, cannot and should not be used as other, static media are to convey information. The Internet is a new, unmatched medium with properties of "emergence" (information exclusively residing in context) and "evolvability" (information that can change and take on new meaning); its potential information modules are multidimensional and take on different meaning with different users and network associations. The Internet's content is constantly changing, updating, and growing. Consequently, we must apply new methods of research, communication, teaching, and learning in order to benefit and gain from its remarkable advantages. In chapter 5, I suggest how and which new methods we might begin to apply. We will see that solutions to particular problems continually faced by teachers and researchers (such as dilemmas within the state standards and core curriculum movements and limitations of methodologies for reconstructing historical patterns) will not come from applications of the same methods and approaches strictly associated with static media. Rather, these solutions will flow from the new, dynamic structure and function of the Internet.

Because the Internet is inherently dynamic, far-reaching, and interdisciplinary, its use and application are not compatible with textbooks, static learning media, or conventional teaching methodologies. As an alternative, the Internet offers the element of discovery learning to students of science unlike any other learning tool. In chapter 6, I demonstrate ways in which the Internet can be used as a powerful tool for discovery- and problem-based learning through global collaborations, whiteboard technologies, and context-based hypertext.

I will not outline detailed examples of discovery- and problem-based learning techniques, lessons, and exercises; there is already a wealth of information available from other experienced educators on these subjects areas, some of whom I will make reference to throughout this book. My intent is not to come across as simply "reinventing the wheel" when it comes to educational pedagogies and applications. Rather, what I see as new is the Internet as an unmatched tool for putting the theories that we long thought would work in scientific education into real and effective practice. In addition, my aim is to show that the goals and practices of discovery- and problem-based learning are greatly enhanced by Internet technology and that their future development and application cannot be fully achieved outside an online arena.

I believe that students of science will study, learn, and discover at their own pace and along their own lines of interest through interdisciplinary efforts that are particularly enhanced by the structure and content of the Internet. In doing so, students will reach an attainable and comprehensive learning standard, one that requires students to use information and knowledge to discover, hypothesize, theorize, and create. This universal standard of learning, encouraged by Internet structure and content, will focus on critical thinking and basic problem-solving techniques that may be mastered and applied across all scientific disciplines.

In chapter 7, I suggest how both teachers and the Internet might function together as learning facilitators for students as well as some approaches and guidelines for mentors and teachers to take that chiefly incorporate elements of Internet discovery learning into courses of study and research. With all that is known throughout human history becoming accessible to all of us via the Internet, educators, more than ever, will need to guide students through the explorations of content.

Rather than representing direct sources of knowledge, teachers will instead be direct participants in the exploration of knowledge, encouraging students to discover facts and helping them identify valid information as they click through hypertext, forming their own semantic pathways and online information modules.

Humans have always been fascinated with opportunities for adventure and exploration. Like the depths of the oceans, the span of the continents, or the vastness of the universe, the Internet also offers a unknown domain to be explored, awed, and appreciated. To me, nothing is more exciting than embarking on boundless journeys through the frontiers of cyberspace. I must admit, as I sit here writing this chapter, that I keep hearing William Shatner's voice uttering, "Space—the final frontier." This popular phrase, which appeals to one's ultimate sense of exploration and curiosity, no longer applies solely to the unreachable boundaries of cosmic space—it applies to cyberspace as well. As we come to recognize the Internet as an entirely new tool for knowledge exploration, discovery learning, problem solving, communication, and research, we cannot help but be amazed by its perpetual evolution and infinite expansion. It is my hope that after reading this book you will begin your own personal explorations of the Internet, navigating your own unique sectors of inquiry and information exchange. By doing so, you will instantly become an everlasting participant in the development and advancement of human knowledge.

2

DYNAMIC CONTENT

As you will see throughout this book, I may use the terms "Internet" and "World Wide Web" interchangeably. I do not mean to confuse any significant distinctions that can be made between the two designations, distinctions, for example, that might separate the actual patterns or connective structure of hypertexted web pages from the processes by which the patterns are generated. Instead, I wish to convey an overall concept of online information storage and exchange. But first, let's recognize the authentic meaning of the Internet as it has been defined by one of its creators, Tim Berners-Lee (2000), according to whom the Internet is

> a global network of networks through which computers communicate by sending information in packets. Each network consists of computers connected by cables or wireless links. (p. 223)

He further defines the World Wide Web as

> the set of all information accessible using computers and networking, each unit of information identified by a URL. (p. 239)

As such, the World Wide Web consists of information (in the form of web pages identified by Uniform Resource Locators [URLs], or web addresses) that is shared between and stored by a network of computers that collectively make up the Internet. In other words, the World Wide Web is the information itself; the Internet, by contrast, is the network that stores the information and allows it to be shared and exchanged between users. I must admit that I have trouble distinguishing between the information itself and its method of access when it comes to the everyday use of online resources. Some might argue that this is sloppy thinking, but as I see it, the World Wide Web and the Internet are interdependent, together forming an emergent, new medium of information exchange.

Information that resides on the Internet is truly dynamic, continually transforming among "sectors of points" (information modules; see chapter 1) that are particularly designed by a user's informational needs and intent. As may be obvious to most of us, any individual website consisting of written or graphical content on the Internet is capable of real-time change in meaning or format according to its author's objective. No other communication medium allows for this type of instantaneous change in the purpose, delivery, and subsequent retrieval of discrete information. Only through the Internet can written information, commonly accompanied and enhanced by audible and visual data, be instantaneously updated and simultaneously dispersed across the globe.

What does all this mean for scientific learning? In my view, it means absolutely everything. Dynamic, real-time content is critical to the advancement of collaborative research in all scientific disciplines. New data are constantly being discovered and synthesized. The steady dispersal of new data can accelerate the reproduction and corroboration of results, culminating in the falsification of old hypotheses and the postulation of new ones as well as the reevaluation of theories and the recognition and dismantling of long-established misconceptions. Alternatively, the lack of access to updated information may hinder and delude related studies being conducted across the globe. We will examine the far-reaching capabilities of corroborative research via the Internet in chapter 4. Here, let us first explore ways in which Internet content might be accessed and how such access actually contributes to its dynamic nature.

Internet browser software created by Berners-Lee in 1990 provided computer users with a hypertext environment (known as the World Wide Web) in which to explore electronic information. "Hypertext" refers to the ability to link together different types of content, in the form of web pages, by "point-and-click" movement throughout the World Wide Web environment. This point-and-click movement is also referred to as "hyperlinking" and characterizes a fundamentally new way of organizing information. Until now, information was serial, and taxonomies were primarily hierarchical. With hypertext, however, "anything can be linked to anything" (Berners-Lee, Hendler, & Lassila, 2002, p. 26). As a result, information on the Internet is completely decentralized, and the semantic pathways (reflected in what I refer to as online information modules; see chapter 1) that are constructed by individual users are often multidimensional.

Critics quickly pointed out in the early stages of the development of the World Wide Web that without a central database or treelike structure, the World Wide Web could not typify a well-organized library of information (Berners-Lee et al., 2002). Indeed, I argue that the World Wide Web is not a "library" of information in any traditional sense. As such, conventional methods of organizing static media, such as branching hierarchies or the Dewey decimal system, will not succeed in organizing hypertext; one can never be sure of finding everything or even anything relevant to a query on the Internet in such a fashion. Instead, new methods of information storage and retrieval must be applied. Unlike libraries and serial databases, the Internet is making far greater amounts of information continually available to users through its dynamic and decentralized structure, and search engines are just beginning to produce remarkably complete indices of much of the Internet's content. As we will see, the semantic pathways created by individual users (and soon to be by machines) are providing new ways that are needed to organize and extract information from the Internet.

Typically, the first step in exploring Internet content involves consulting a search engine. Search engines, such as Google, Altavista, Yahoo, and Northern Light, are tools that scan the Internet for information about a particular topic, key word, or phrase and prioritize websites in order of popularity, importance, or relevance to the designated query. There are several sources that explain the underlying mechanisms that make search engines work ("Clever" Project Members, 2002; Lewin, 2001; Lynch, 2002). I do not present the details here but rather elaborate on the dynamic process of hypertext exploration and its relevance to a new application of search engines called the "semantic web" (Alesso & Smith, 2002; Berners-Lee et al., 2002).

One of the most powerful and unique aspects of Internet content is its ability to exist within multiple contexts. All information on the Internet is destined to become context based by way of its access: the process of hyperlinking and the formation of online information modules. As a user explores hypertext through point-and-click movement, associations between web pages and the information they contain are intuitively made, as is an expression of an individual user's thoughts and fact assimilations during the particular search process. The way in which these associations are made is strikingly reminiscent of some of our own methods of learning. For example, there have been several psychological studies of word associations and network simulations suggesting that we carry semantic maps in our brains: representations of words and their meanings are stored in the cortex in a network-like structure (Jensen, 1998; Spitzer, 1999). Within such a network, there exists nodes and branches. Nodes are actual words and their structure, whereas branches connect nodes with nearby nodes, thus placing a word within a larger context of meaning. Within the network-like structure of a semantic map, the meaning of a word is represented both by a node and by the relation of the node with other nearby nodes (see figure 2.1).

In the practice of learning or teaching language, we are quick to discover that the meaning of words depends on the context in which they occur. Mel Levine, author of *A Mind at a Time* (2002), effectively describes how students fit words into semantic categories while learning to understand their meaning:

> Rather than simply memorizing its definition, they note that the word [*altruism*] represents a form of kindness and that it is related to charity, helping and caring for others, and self-sacrifice. They may also sense that altruism is the opposite of selfishness, miserliness, and insensitivity to others. . . . They understand shades of meanings and they are adept at comparing words to other words, to words that are opposites, to words that are similar but not quite the same, to words that are synonyms. (pp. 135–136)

Students place words into semantic categories and subcategories and are able to cross-index terms such that a single word may be thought of as a process, concept, or state of being (figure 2.2). Our abilities to understand and assimilate information, which are highly diverse among groups of learners, rely on the construction of semantic networks that place the information within the needed context. We acquire new

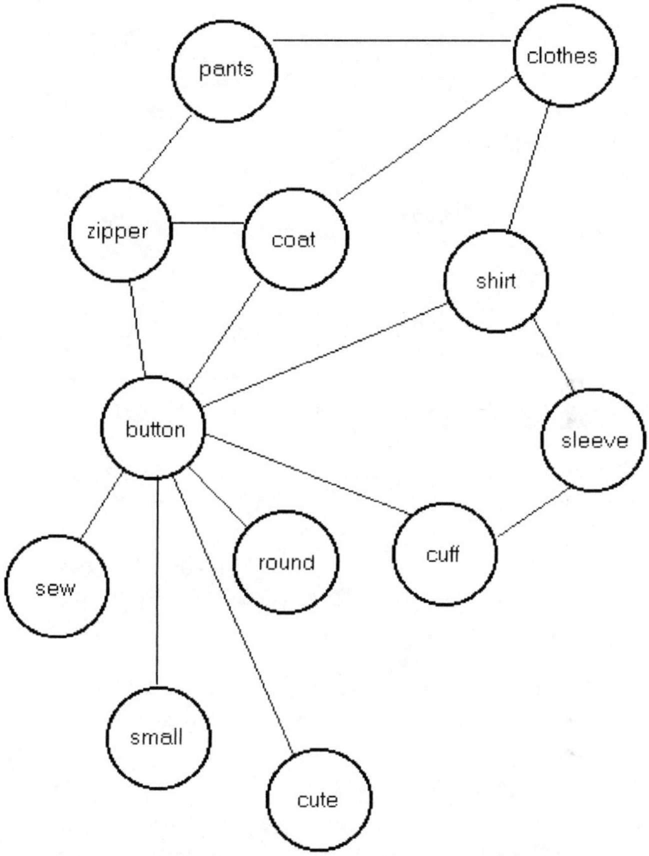

2.1 Portion of a semantic network of word associations. Words are located at nodes within the network, and lines represent the relation of nodes with other nearby nodes.

meaning by comparing new words or information to terms and concepts that we already know. Hyperlinking, unlike any other form of information processing, provides a method for constructing our own "out-of-mind" semantic networks, or pathways, of information that are characteristic of our individual strengths, abilities, interests, and current knowledge, thus enhancing our ability to synthesize and expand our knowledge base.

To play around with this idea, I asked a ninth-grade class to research information about the human digestive system using only Internet

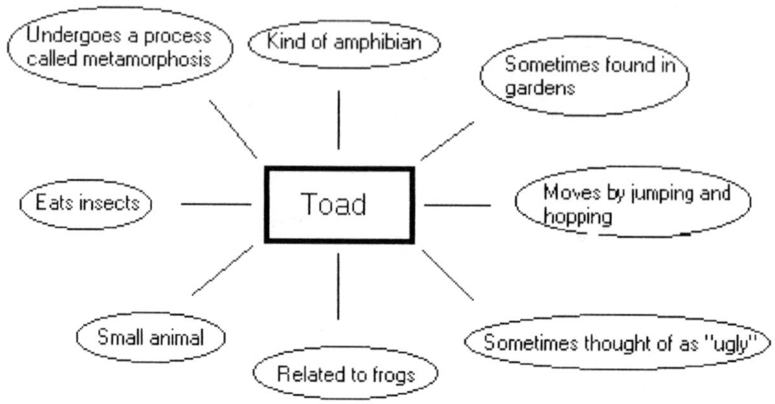

2.2 A semantic map of meanings associated with the word *toad* (adapted from Levine, 2002, p. 136).

sources. I introduced them to the concept of the digestive tube, the organs that constitute it, and some of their basic functions. I then asked each student, on his or her own, to find other facts and information that can help us better and more fully understand the structure and function of the digestive system. Each student would present the class with an informal, oral synopsis of his or her findings.

The immediate response from the class, entirely and overwhelmingly, revealed their intense concern over where to start looking, what exactly to look for, and how to organize and present the information. The only guidelines I offered included important criteria to follow when trying to identify high-quality, reliable sites on the Internet (a topic that I mention in chapter 7). I also required that they save and document all URLs of the websites at which they stopped, sequentially, during their online journeys. My main reason for this was to be sure that the students were accessing quality sites and that they performed an active exploration rather than just stopping at the first site that their search engines generated.

More students than not came back with regurgitated information that was unsynthesized, taken strictly from a website or two, and repeatedly covered the basics of structure and function that we had already discussed as a class. I was not surprised. Most of these students had never before been asked to complete an assignment that was not to be graded and that asked only that they pursue their own learning objectives. (I address this point further in chapter 6.) But a few students did come back with reports, unbeknownst to them, of the information modules that they had constructed while clicking through hypertext. I highlight four of these student's findings in the next section. Although I cannot fully reconstruct the actual structure of their information modules or the relationships among the concepts and facts that make up those modules, I can list for you the different topics and ideas that arose in each exploratory search. All four of the students began their explorations gaining a better understanding of the structure and function of digestive organs. Then the meaning of the digestive system took on distinct relevance for each student:

Student 1

- Structure and function of
  - Human digestive organs
  - Accessory organs
  - Digestive enzymes
  - Bacterial populations in the intestines
- What happens when something goes wrong with the digestive system?
- Symptoms and causes of
  - Ulcers
  - Food poisoning
  - Gallstones
  - Dysentery
  - Parasitic infections

Student 2

- Structure and function of human digestive organs
- Relationships between the processes of ingestion and digestion
- Types of nutrients
  - Humans are heterotrophs
- Types of heterotrophs
  - Carnivores
  - Insectivores

    - Omnivores
    - Scavengers
    - Herbivores
- Food pyramids and webs

Student 3

- Structure and function of
    - Human digestive organs
    - Accessory organs
    - Digestive enzymes
- Hydrolysis of macromolecules
- Absorption of nutrients
- Processes of cellular respiration
- Relationships among processes of digestion, circulation, respiration, and hormone production

Student 4

- Structure and function of human digestive organs
- Embryonic origin and development of digestive tube
- Digestive tube homologies within animal kingdom
- Alternate digestive/ingestive structures and processes within animal kingdom

These topics help describe sectors or clusters of points within the semantic maps that were constructed by each individual student during a single exploratory search on the Internet. Although the initial topic was the same for all—the structure and function of the human digestive system—its meaning for each student came from what they perceived as relevant to their own understanding of the significance of digestion as it relates to the following:

1. Disease
2. Ecological niches and food webs
3. The function of other physiological systems of the human body
4. The development and evolution of animals

In chapter 6, I attempt to express these ideas in the form of Venn diagrams, which may better illustrate the resultant knowledge formulated by the process of building online information modules.

As a science educator looks at the previous list of topics, he or she would immediately notice that those topics are already included in most state biology/general science curriculum standards. But what is important to note is that these topics were not presented *to* the students; rather, the students came up with them on their own, based on their own interests and queries, and constructed the significance of the topics on their own.

Now, you might ask, Could research in a library, using traditional sources such as books, encyclopedias, and periodicals, lead to the same result, that being students constructing their own meaning of the human digestive system while at the same time expanding their overall knowledge base of the system? For more experienced students, I think that the answer is yes. However, (1) the amount and type of materials and information sources is more limited in a library, (2) the entire research process would take much longer, and (3) the method of making quick connections between concepts and topics (that which immediately overlaps with the mind's thought processes and semantic map construction) would not be available to a student working in an offline environment. With respect to all these points, information important to the student's understanding may be lost or never encountered while accessing traditional sources outside an online arena.

Given this outlook, I do not suggest that an Internet query be the end-all of a scientific research project. But it is by far the best starting point for constructing meaning and knowledge about a subject that can and should be supplemented with other forms of media (that is, while these media forms still exist outside the Internet). Subsequent and equally important aspects of research require students to pose and test hypotheses to further expand their understanding. As I show in later chapters, the Internet can greatly aid in this endeavor as well.

Although individuals can create their own online semantic maps through hyperlinking, a new development aims to automate part of this process. The semantic web, which continues to be conceptualized by Tim Berners-Lee and others (2002), is not a different web but rather an extension of the current one, in which information is given well-defined meaning, better enabling computers to process and "understand" the web pages that they merely display at present. As Berners-Lee envisions,

> Computers will find the meaning of semantic data by following hyperlinks to definitions of key terms and rules for reasoning about them logically (p. 25). . . . Like the Internet the semantic web will be as decentralized as possible. . . . For it to function, computers must have access to structured collections of information and sets of inference rules that they can use to conduct automated reasoning. . . . Adding logic to the web—the means to use rules to make inferences, choose courses of action and answer questions—is the task before the semantic web. (p. 26)

In a similar sense, Alesso and Smith (2002) describe their vision of an intelligent wireless web as

> a network that provides anytime, anywhere access to information resources with efficient user interfaces and applications that learn and thereby provide increasingly useful services whenever and wherever we need them. . . . For the Web to learn, it requires learning algorithms and mechanisms for self-organization of a hypertext network. It needs to develop algorithms that would allow the Web to autonomously change its structure and organize the knowledge it contains by "learning" the ideas and preferences of its users. (pp. 22–23)

Alesso and Smith (2002) further articulate that the World Wide Web was originally designed as a space to display information, with the goal that it should be used not only as a communication medium for humans but for interactions between machines as well. They suggest that one of the major obstacles for machine communication via the World Wide Web has been the fact that most of the information there is designed for human consumption, and

> even if it [information] was derived from a database with meanings for its database elements, the structure of the data is not evident to an autonomous agent browsing the Web. Leaving aside the artificial intelligence problem of training devices to behave like people, the Semantic Web approach instead develops languages for expressing information in a form that a device can process. (p. 213)

Does all this mean that computers will soon do our thinking for us as we search for information on the Internet? In my view, the construction of our own individual human semantic pathways cannot be automated. We all bring our own uniqueness to Internet explorations and interpretations of content. Although semantic search engines may automate part of the association process and help computers communicate with each other, profound explorations of Internet content and the formulation of human knowledge are achieved through our own individual attempts at information retrieval and exchange. Each time a user initiates point-and-click movement through hypertext, regardless of the number of repeated queries involving the same key word or topic, new semantic pathways (reflected in information modules) are created, leading to new underlying sources of knowledge. As such, Internet content is infinitely dynamic, continually allowing for the construction of boundless semantic pathways of online information by way of its users' access.

This practice may seem chaotic and intractable such that a single result or solution to a query may never be repeated. Indeed, Internet query results are dynamic. But I argue that their dynamic nature in fact augments our interpretations of information by providing multiple and distinct contexts that may elucidate problematic conditions existing not only at the microscopic level of inquiry, involving an isolated and specific set of investigational conditions, but at the macroscopic level as well, including an understanding of the structure and function of interrelated or transcending systems. By collaborating and placing information in several different contexts, each according to our own expertise, situation, and perspective, we gain powerful insight into the significance, implications, and achievability of a macroscopic view of scientific knowledge. This context-based, macroscopic view of knowledge, its attainment through hyperlinking, and its importance to scientific inquiry make up the focus of the next chapter.

3

LIMITATIONS OF REDUCTIONISM

As we learn new concepts and skills, an applied strategy that works for many of us involves breaking down new information into separate tasks or parts. Understanding the underlying structure and significance of the individual parts of an idea and how they build on one another to constitute a more complex notion can help us acquire new knowledge that conveys to us a comprehensive and overall meaning of the idea. As students, we engage in the process of "breaking down information" constantly while learning to read, expanding our vocabulary or manipulating mathematical equations. But in order to effectively acquire new knowledge, we must do more than simply break down complex wholes into smaller, more tractable subparts; we must also elucidate the relationships among the parts that make up the whole.

This latter process is not readily obvious to some of us. It may seem more intuitive to assume that the whole of anything equals the sum of its parts. In this event, we tend to focus on identifying processes that are, at best, limiting in their potential to elucidate how patterns in nature, human behavior, or even knowledge acquisition are formed. The reality is that simply summing together the individual effects and properties of the subparts is not the sole course of action underlying the construction and assimilation of knowledge. Moreover, I would argue that "additive effects" of multiple parts are not representative of how nature itself is structured, how it evolves, or how it should be analyzed. Let's see how the limiting pedagogies of additive effects have typically come into play and why it might be best to move beyond their scope as we strive to expand our understanding of the world and our place within it.

In science, there are several different perspectives and approaches one may take while studying natural phenomena. I personally have come to recognize these as conforming to either a "microscopic" or a "macroscopic" view of nature. Analysis at the microscopic level requires the reduction or breakdown of natural patterns into identifiable parts. Many would argue that subsequent focus on the structure and function of these parts is essential to gaining an understanding of the whole pattern or working system that the parts constitute at a macroscopic level of analysis. Traditionally, investigations under various fields of view, micro- or macroscopic, are considered equally essential for explaining behaviors at distinct levels of biological organization. And, not surprisingly, different questions arise when the scope of our investigative fields changes.

The questions that arise from different fields of view ensue our perspectives about the natural world. In turn, these perspectives might be encapsulated by two distinct, competing questions: is the whole understood by knowing its parts, or are the parts understood by knowing the whole? This dilemma has been at the forefront of my mind for some time, even as I am engaged in ordinary tasks. For instance, as I was casually walking through the grounds of Wave Hill last week, a most beautiful botanical garden in the west Bronx, I came across a Copper Beech tree of amazing stature and morphology. At first, my attention was drawn to its small, fiery-red leaves. While observing the leaves with my trusty hand lens (what field biologist leaves home without it?), the reduced field of view triggered questions in my mind regarding the structure and function of specialized leaf cells, the types of pigments, and the mechanisms of light absorption, the processes of gas exchange, the biochemical pathways of photosynthesis, and the differentiation patterns of leaf ontogeny.

As I crouched under the canopy of the first set of branches and looked up toward the top of the tree, I was awed by its overall height and girth, its beautiful pale gray bark, and its magnificent array of branches. My thoughts shifted from inquiries about the subparts and processes of leaves to inquiries about the overall organism that was before me: How is its branching pattern regulated? What might be adaptive features of its leaf morphology, growth habit, or reproductive processes? This simple exercise of observation emphasized for me the way in which the changing of vantage points and the scope and depth of our analytical foci may elucidate questions that actually transcend various systematic levels of organization. Although traditionally segregated, I believe that both micro- and macroscopic studies can be advanced and expanded by moving from a primarily reductionist approach to a context-based approach to scientific inquiry that encompasses several different levels of systemic organization.

Before I explain what I mean by a context-based approach, let me first offer one example of a prevalent scientific study that instead helps to clearly demonstrate an exclusive reductionist type of analytical approach, the Human Genome Project (Human Genome Project Information, 2003), the initial goal of which was to identify all the genes (the parts) that make up human DNA (the whole). Sequencing the entire genome involved collaborative work from various research institutions, aggregated by the National Institutes of Health and the Department of Energy: Washington University (St. Louis), the Baylor College of Medicine (Houston), the Joint Genome Institute (Walnut Hills, California), and the Wellcome Trust medical charity (Cambridge, England). These five sequencing centers became known as the "G5" and collected 80 percent of the data (Micklos, Freyer, & Crotty, 2003). Fifteen other major sequencing centers in the United States, Europe, and Asia worked together in sequencing other parts of the genome.

At first, the project aimed to develop chromosome maps and markers needed to identify genes involved in diseases and to provide anchor points for specifying the location of DNA sequences (Micklos, Freyer, & Crotty, 2003). Proponents of the project readily point out that naming and identifying all genes in the human genome is a giant step forward in understanding how our cells work and how particular gene activities contribute to disease. But this reductionist approach to scientific knowledge has, in turn, giant limitations. As Micklos et al. (2003) explain,

> The huge amounts of data generated by the high-throughput methods to sequence genomes and to analyze gene expression may lead us to believe that the work of biology is nearly done. Nothing could be further from the truth. By all accounts, the functional analysis of genes discovered by the human and other genome projects will be much more painstaking than their sequencing. (p. 213)

> Now that scientists have turned over a catalog of tens of thousands of human genes, the challenge remains to understand how these genes work together. Although every nucleated cell (except germ cells) possesses the same set of genes, different cells express different sets of genes at different levels. Understanding differences in gene expression is vital to understanding the development and the function of different cell types. (p. 209)

Other studies that initially or entirely invoke a reductionist approach are represented in all scientific disciplines, from physics to chemistry, from ecology to molecular biology, and from taxonomy to physiology. I would argue to an extent that the reductionist approach has grand validity and importance. Indeed, much is gained from observing details and identifying individual components of the natural world. For example, how might we classify members of the plant kingdom, elucidate their evolutionary relationships, and understand their structural and functional makeup if we do not know the detailed biochemical pathways of photosynthesis or the specialized cell types that make up their reproductive or vascular tissues?

However, what is missing from a reduced, "piece-by-piece" approach to studying the natural world is an understanding of the context in which the details reside, how components relate to one another, and how such relationships result in emergent structures and levels of complex organization (Bar-Yam, 1997, 2000; Bass and Emmeche, 1997; Clark, 1996; Oltvai & Barabasi, 2002; Wolfram, 2002) that exhibit distinct characteristics, functions, and capabilities. Relating back to the Human Genome Project, I am reminded of a statement made by the evolutionary biologist Ernst Mayr (2001): "DNA itself provides the building material of a new organism, but it is only a blueprint (raw information) for making the proteins of the phenotype" (p. 272). I interpret this statement as indicating that our explanations of an organism's phenotypic development cannot be reduced solely to analyses of gene activity. Instead, an understanding of the context in which genotypes and phenotypes relate, indicated by the seemingly enigmatic processes of ontogeny, is needed.

My first real, "in-your-face" encounter with reductionism occurred when I was in graduate school. I was conversing with a professor of mine in the laboratory when a call came down from one of the department's plant biochemists. He said that he had a graduate student working with him to understand the biochemical effects a gene was having on the cortex development in stems. I do not remember the gene, the biochemical pathway under investigation, or the plant (although I think they were studying *Arabidopsis*). But what is forever stamped on my brain is the biochemist's request to have the student quickly come down to our lab and find out what a cortex is.

For those of you who are not familiar with plant biology, the cortex is an internal tissue of a stem or root bounded externally by the epidermis and internally by the xylem and phloem. When the student came down, he had never once examined the anatomy of a flowering plant before and knew practically nothing about his organism's histology, anatomy, or morphology. My professor boldly asked the student, "Your work seems interesting, but how can you really understand it if you don't even know your organism?" He seemed very perplexed by this question and replied simply that his research focused on studying gene activity. My professor then recommended that he take her plant anatomy course next term.

My point in bringing up this story is that sometimes one can get so focused on a reduced level of inquiry that one can miss the overall big picture (that is, the significance and implications) of the work and its multifaceted contributions to science. In addition, let me clearly point out that I am not against the work of plant biochemists. Their work is immensely valuable, and its relevance can truly extend beyond the parts of biochemical systems and give us a better understanding of how organisms live and adapt.

On the website for the New England Complex Systems Institute, Yaneer Bar-Yam (2000) skillfully explains that when parts of a system are related to each other, we often talk about them as a larger network of interacting components and that "emergence is what parts of a system do together that they would not do by themselves" (p. 1). According to this definition of emergence, system analysts aim

> to understand how collective properties arise from the properties of parts. More generally, [emergence] refers to how behavior at a larger scale of the system arises from the detailed structure, behavior, and relationships on a finer scale. (p. 1)

Bar-Yam (2000) further describes the study of emergence as involving the movement between different vantage points:

> When we think about emergence . . . we see the trees and the forest at the same time. We see the way the trees and the forest are related to each other. To see in both these views we have to be able to see details, but also ignore details. The trick is to know which of the many details we see in the trees are important to know when we see the forest. . . . In conventional views the observer considers either the trees or the forest. Those who consider the trees consider the details to be essential and do not see the patterns that arise when considering trees in the context of the forest. Those who consider the forest do not see the details. When one can shift back and forth between seeing the trees and the forest one also sees which aspects of the trees are relevant to the description of the forest. (p. 1)

To illustrate this idea of emergence further, one that I think conveys a context-based approach to inquiry and analysis, let us consider a few examples of relatively large and complex structures that can be broken down into smaller, simpler parts.

The Human Body. The structure, function, and behavior of the human body is the culmination of numerous biochemical actions dictated by proteins. Proteins consist of a chained sequence of molecular building blocks called amino acids. Instructions for building and chaining together particular amino acid sequences are coded by genes, which are in turn specified by nucleotide sequences consisting of the nucleic acid molecule known as DNA (figures 3.1 and 3.2). Individual nucleotides alone do not specify genes. Only when they are arranged within a specific context (or sequence) along the DNA molecule is information, in the form of genes, generated. Furthermore, the relationship between proteins and their ultimate encoders (genes) is rarely direct and straightforward. The "one gene, one protein" relationship is not commonly the case when it comes to phenotypic expression. Many proteins require the expression of more than one gene. Alternately, some genes express more than one protein. Moreover, the expressions of some genes are contingent on the expressions of others or may be altered by specific environmental conditions. Thus, the whole, in this case the developed human body, is not simply the direct sum of all its parts identified at a reduced, molecular level. Rather, the relationships between the parts (nucleotides, genes, amino acids, and proteins)—how they interact

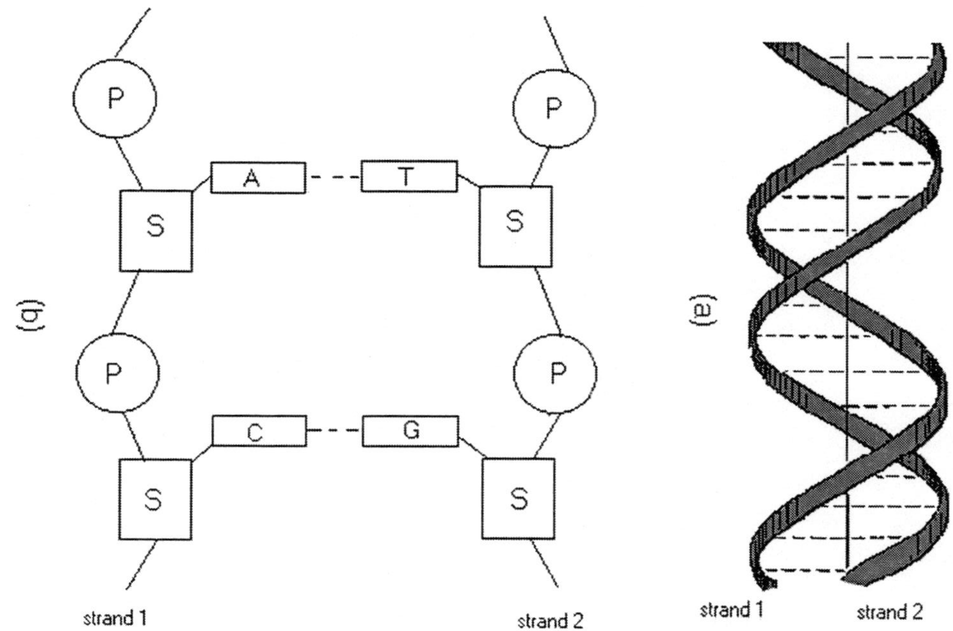

3.1 A simplified illustration of the structure of DNA: (a) two strands of molecules, each consisting of nucleotide units, arranged in a double helix, and (b) two sequential nucleotide units, each consisting of a phosphate group (P), five-carbon sugar (S), and a nitrogenous base (A, T, C, or G). Pairs of nucleotides in the DNA double helix are held together by hydrogen bonds, indicated by the dashed lines in both (a) and (b).

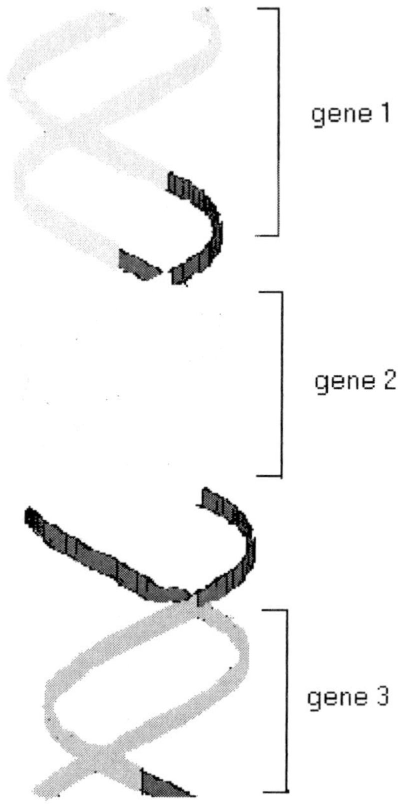

3.2 Particular sequences of nucleotides in DNA specify individual genes.

at and across various systematic levels as well as with the environment—are critical to our understanding of how they culminate in the structure, function, and behavior of the human body.

Written Language. The structure and meaning of a written document come from the cumulative influence of the individual sentences that make up that document. Alone, these sentences do not provide the reader with the full scope and significance of the written work. But together they constitute and reveal an interdependent scheme that requires interpretation by the reader. Likewise, the combination of individual words (rather than the words alone) in a sentence, along with grammatical notations and punctuation, provides the reader with the overall meaning of the sentence. It is the context of these words, their syntax, and how they relate to one another and other linguistic notations that are key to our understanding the meaning of the sentence (figure 3.3). If we are to break words down into more reduced subparts, we of course find that letters make up individual words. However, like nucleotides, letters must be placed within a specific context, or sequence, in order to form a word and convey a particular meaning or meanings. Interestingly, in many written languages, there are words that may have identical letter sequences (that is, spelling structures) yet have completely different connotations. The only way to understand their meanings is to read them within the *context* of other words in a sentence. For example, the word "work" has a different meaning in "I went to work today" than in "I work in the laboratory on Tuesdays and Thursdays."

Online Information Modules. As I have defined them, online information modules are user-specific semantic pathways built by point-and-click movements through the hypertext (for example, links) of web pages. The existence of these modules is based solely on the context of individual web page links, reflective of a user's semantic network (see chapter 2). They are the "out-of-mind" semantic maps of the user. They cannot be reconstructed but, rather, only communicated between information exchangers (although Judy Breck's [2003b] illustration of a Jumpstart Pavillion may help you visualize the concept). The user gains knowledge and meaning by interpreting how the linked information is related rather than from static or isolated information provided by the individual web pages themselves.

In all three previously mentioned examples, as the parts interact, new levels of design arise, distinguished by an increase in the number of interrelated contingencies, that is, an increase in context-based information. We must recognize that the context of parts is as important as the

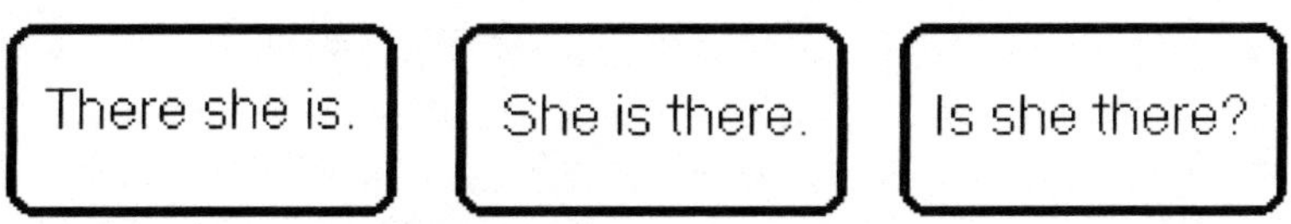

3.3 In English, syntax (word order) is important to changing the meaning of a sentence. Note that the words *she*, *is*, and *there* in the boxes can be combined in three different ways to alter meaning (adapted from O'Neil, 2002, p. 1).

structure and function of the parts themselves to our overall interpretation and understanding of the information that they provide. While conceptualizing the World Wide Web, Tim Berners-Lee (2000) expressed the following:

> I liked the idea that a piece of information is really defined only by what it's related to, and how it's related. There is really little else to meaning. The structure is everything. There are billions of neurons in our brains, but what are neurons? Just cells. The brain has no knowledge until connections are made between neurons. All that we know, all that we are, comes from the way our neurons are connected. (p. 12)

If we solely analyze parts separate from their associated parts and hence apart from their relevant context, significant information is lost, and our explanations of phenomena and communication of knowledge may remain incomplete and problematic. In terms of our understanding of the vast scientific mysteries of evolution, for example, the evolutionary biologist Ernst Mayr (2001) suggests that

> evolution is not a matter of changes in single genes; evolution consists of the change of entire genotypes. . . . Such frequent, in fact universal, interactions among genes are of decisive importance for the fitness of individuals and for the effects of [natural] selection. (p. 272)

An ordinary example that may help convey the significance of a context-based approach to inquiry comes from considering the structure and function of a key. As Bar-Yam (2000) describes,

> A key has a particular structure. Describing its structure is not enough to tell someone that it can open a door. We have to know both the structure of the key, and of the lock. (p. 2)

Thus, we may come to understand the structure of the key by studying it in isolation of its lock counterpart, but we can never fully understand its function unless we analyze its structure in relation to that of the lock and how these two structures function together as a "locking" system.

The moral of the story that I have been presenting here is that we need to seek out the "big picture" of our investigations, whether they are new or revisited. Viewing problems macroscopically can alter the scope and implications of any form of inquiry by taking into account the relevance of other data (even from disciplines that may seem irrelevant at the microscopic level of inquiry), recognizing other large-scale patterns and processes, and seeking common explanations for these observed events. As others have suggested, searching for patterns that arise from the relationships among details and their contingencies within a system can be considered an essential approach for understanding intelligence (Bentley, 2001; Dyson, 1997), the evolution and expansion of life (Bar-Yam, 1997; Dyson, 1997), phenotypic development (Gerhart & Kirschner, 1997), the structure and function of neural networks (Bar-Yam, 1997; Spitzer, 1999), civilizations (Bar-Yam, 1997), and social groupings (Johnson, 2001).

Some new and intriguing, context-based approaches to macroscopic studies of complex systems involve a constant shifting of our analytical foci via mathematical analyses of the existence of spontaneous order in nature and the universe. For example, in *The Origins of Order*, Stuart Kaufman (1993) suggests that the origin and evolution of life are actually lawlike events governed by principles of self-organization. In *A New Kind of Science*, Stephen Wolfram (2002) suggests that basic forms of behavior in physical, biological, and other systems occur over and over again, almost independent of underlying details, and are guided by universal rules and computations. Alternatively, Steven Strogatz (2003) describes a natural phenomenon, called sync, in which many cycles occur simultaneously at several levels of biological organization:

> At the lowest, most microscopic level, the cells within a particular organ are mutually synchronized; their chemical and electrical rhythms vary in lockstep. At the next level, synchrony occurs between the various organs, in the sense that they all keep to the same period, even though the cells have differentiated into disparate types. . . . The sync is in the sense of period matching, keeping the same beat, just as musicians keep the beat in their heads even when they are quietly awaiting their turn to play. Finally, the third level of synchrony is that between our bodies and the world around us. Under normal conditions of living on a regular schedule, seeing sunlight, sleeping at night and so on, the entire body is synchronized to the 24-hour day, driven mainly by the cycle of light and darkness. (p. 72)

Strogatz suggests that the circadian rhythms of organisms represent a process of external synchronization in which an organism falls in step with the environment. Trying to decipher the mathematical rules for human circadian rhythms is one important application of the new science of sync:

> We are trying to learn the rules of the circadian symphony. . . . We know such rules exist because we can see their manifestations at the larger scale, in the behavior of a whole, integrated human body: in our daily rhythms of sleep and wakefulness, hormone fluctuations, digestion, alertness, dexterity, and cognitive performance. At this level, scientists have recently discovered cryptic regularities in the timing of human sleep-wake cycles and other circadian rhythms, even though the microscopic basis for these laws remain enigmatic . . . we now know that human circadian rhythms obey their own kinds of laws, though we remain in the dark about their fundamental biochemical basis. (p. 73)

As these new approaches to scientific inquiry suggest, the behavior of a whole system is critical to our understanding of the underlying mechanisms responsible for observable patterns in nature. Furthermore, detailed knowledge of the parts of a whole should not necessarily be the foci for elucidating an array of common laws that may dictate complex behaviors.

Because the acquisition of knowledge from the Internet demands placing information within large-scale contexts, I believe its common and continual use will ultimately change the way in which we learn, study, hypothesize, and conduct scientific research. During my years as a graduate student, I conscientiously observed the research agendas of several scientists, both at my own academic institution as well as at others (national and international), while trying to develop my own scientific approach and practices. I was surprised to find that few ventured out of their "study group," whether it entailed a specific tissue type, gene sequence, population, organism, or community of organisms. In fact, these researchers seemed to ask the same types of questions over and over again at the same level of inquiry and were critical of those seeking explanations of larger-scale patterns in nature, arguing that questions at such a level were untestable. However, I thought (and still do) that having a macroscopic perspective was interesting and important as well as scientific. The problem, rather, remains that the theories, techniques, and approaches being applied at the reductionist level (those that we are so familiar with and trained to use) are not necessarily those that should be applied at the macroscopic level. In my view, the major challenge that scientists continue to face (but that many overlook) is the development of hypotheses, tests, and theories that encompass several emergent levels of pattern and process development.

To examine both the patterns and the processes of nature, we must expand our focus from the individual parts of a whole to include the relationships between parts of a system and its environment. To emphasize this point in a very simplistic way, I always think back to the pioneering experiments and explanations of the patterns of heredity made by Gregor Mendel in the late nineteenth century. As Mendel discovered, it is not the form of heritable factors alone that dictates the expression of particular traits. Rather, what is informative to our understanding of phenotypic expression is how such factors (that is, dominant and recessive alleles) relate to one another at a particular locus. Mendel's law of dominance reflects the mechanism by which the dominant allele is expressed in a diploid organism. Remarkably, Mendel never even "saw" the heritable factors (now known as alleles) that represented the foundations of his work. He instead inferred their existence and function from the context and evidence of his experiments.

Similarly, Watson and Crick (1953) did not "see" the structure of the DNA double helix and all its subunits. Instead, the helical structure was inferred based on X-ray diffraction patterns of the chained monomer configurations. Both Mendel and Watson and Crick demonstrate that a modeling or comparative analysis approach can test hypotheses of a system's emergent function or behavior amidst uncertainty of its structural components.

Another incident that makes me ponder the significance of "relationship" among components (that is, subparts) and their resultant emergent properties concerns a conversation I had recently with some colleagues. We were informally chitchatting about the "secret of life" as it might be viewed scientifically rather than philosophically. I mentioned how a few physiologists from my graduate school days used to argue that the secret of life can be reduced to the existence of electrical and chemical gradients, ostentatiously suggesting that such gradients represent the ultimate foundation on which all life processes are dependent.

Going along with this fun idea, a colleague quickly reduced our foci to the level of submolecular structures and remarked that it is not gradients that represent the "secret of life" but rather the existence of different electrostatic charges. After all, in order to have an electrostatic gradient, there must first be particles with associated net charges. This, of course, is true. But the significance of a particle's charge is understood only in relation to the opposing charge of another particle. What I actually found interesting about this utterly strange conversation is that it is not the existence of different charges alone that is significant to understanding the "secret of life" (whatever that may be—of course we never did come up with it) but how the charges and abundance of particles across a boundary relate to each other in the culmination of a gradient. This assertion leads to endless questions, such as, What constitutes the boundary? Where does it exist? What moves with or against the gradients and how? Not surprisingly, the secret of life cannot be reduced to one particular entity or phenomenon.

Science is not just about finding out what "things" are made of but also about how things and their parts relate to each other and function within particular contexts. I would argue, as I know most others would, that this holistic approach has always been the goal of science. But I do not think that this objective has consistently been made clear to students, especially by their exposure, in conventional classroom curricula, to many of the types of questions asked and tests designed by scientists in various fields. Although we are making significant progress in the elucidation of detail and mastery of technique, in many instances we seem to remain at the reductionist level of inquiry and explanation. Unfortunately, in my opinion, it is these instances that students seem to be made aware of most often. We can identify and tabulate the number of genes in genomes, enzymes in biochemical pathways, cells in ground tissues, or species in ecosystems, but what knowledge and insight are we truly bringing to light when the context of entire biological systems is left out of the range and depth of our analytical foci?

The change in focus from identifying distinct parts or details of a system to also understanding the relationship of these parts is already under way in many disciplines (see the previous discussion) and, in my opinion, might be to a large degree a direct consequence of Internet use for research and communication purposes. For instance, as a student might access the Internet to learn about junk DNA in frogs, through point-and-click hypertext he can also learn about gene sequences in frogs, frog taxonomy, ecology, geographic distributions, evolutionary histories, or the existence, structure, function, and replication of junk DNA in other organisms.

Placing initial queries in a large-scale context may lead to additional questions pertaining to one's investigation. Of course, it is beyond impractical to study a particular phenomenon within the relevant context of all scientific disciplines simultaneously (most likely, the number of relevant contexts is infinite). Studies would take several lifetimes or longer to produce sufficient data. But, as we will see in chapter 4, fast-paced collaborations between scientists of different disciplines around the world are now possible through Internet technology, greatly augmenting the potential and success of large-scale interdisciplinary investigations. Even without such collaborations, placing one's query within a larger context is critical to understanding and accurately conveying the purpose, meaning, scope, and implications of a study.

As I see it, the Internet is the only medium that can help students place their ideas in both a macroscopic and a microscopic context simultaneously, allowing them to move back and forth between such vantage points and decipher information underlying the generation of overall patterns and processes of a system. In my ninth-grade science class, each time I introduce a new topic, I am constantly bombarded with questions from students pertaining to both the details and the general applications and implications of the topic. For example, one week I introduced the topic of triglycerides and unsaturated and saturated fats during discussions we were having on nutrition. My "curriculum objective" was to review the molecular structures of triglycerides, the chemical reactions involved in their synthesis and decomposition, and their general nutritive significance in plants and animals. From the very onset of this triglycerides unit, students bombarded me with questions such as the following:

- Why do some animals make mostly unsaturated fats?
- Why don't all animals have blubber?

- Why are so many people fat?
- Why can't plants get fat [obese] like animals can?
- How did the first cells and phospholipid membranes form?
- How does fat store energy for cells?

Given my curriculum objective, streams of such questions may seem utterly unfocused, distracting, and unproductive. However, the reality is that they are reflective of the semantic pathways being built in students' minds, each with unique relevance and interrelationships among various ideas, facts, and fields of thought as new concepts are introduced and begin undergoing the process of assimilation and comprehension by the individual students. As I introduce and work through various curriculum topics, I cannot answer all their questions or place them within relevant contexts during single class periods or even throughout entire semesters. The Internet, however, can nurture and guide such complex queries, allowing students to swiftly change points of view and pursue individual interests through the creation of their own meaningful information modules.

PART 2

NEW MEDIA

Simple things should be simple. Complex things should be possible.

—Alan Kay, twentieth-century American computer scientist

4

FAR-REACHING CAPABILITIES

The combined use of voice, data, and animation over the Internet creates an otherwise unattainable investigative experience for students and researchers. They can listen to and visualize information as well as simultaneously respond to the information in written or verbal form or through a combination of multiple communication modes. The instantaneous exchange of information between people made possible by the Internet is forever transforming the way in which we research, find, collaborate, and think about information.

During the same time that it takes to collect and analyze scientific data, we can now collaborate with others on it and expand its significance and implications. Scientific collaborations are enhanced by the Internet not just by the rapid pace of information exchange but also by the diverse modes of communication that only it can offer. With no other medium can a researcher simultaneously express or gather information in several written and spoken languages as well as in the form of live animation, photographs, or illustrations.

Why is it so important that we be able to convey information in multiple formats? Throughout my experiences working as a student and being a mentor for other students, the one aspect of learning that has become blatantly obvious to me is its diversity of style and significance among all of us. For me, learning is always enhanced and accelerated when I can view and convey information in various visual formats. I have always been visually oriented as a learner. I understand structures and configurations best when I can see and manipulate them rather than read about them. In addition, new concepts are easier for me to grasp if I can see them conveyed by some sort of visual animation or representation.

I remember from my elementary school days that mathematics came easier for me if I could see and count objects while they were being manipulated rather than interpret the manipulations from notations and expressions. Even when I worked through advanced math in college and graduate school, I found myself picturing in my mind "bunches of apples" when trying to understand certain equations. Reading and writing comprehension tasks were challenges for me as well and still are today to some extent. (I think that my graduate studies advisers aged exponentially when reading drafts of my dissertation.) In many instances, I understand better when I can hear and see someone speaking rather than when I read extensive text. Nonverbal expressions from the eyes and hand gestures actually help me interpret and assimilate new information, and I am usually more effective at conveying information in the form of oral presentations rather than written documents. (But do not let that stop you from reading this book.) My point is that the more conventional means of communication were not my forte, and I am not alone when it comes to expressing this type of educational characteristic. I often wonder now how my early learning might have been influenced by Internet content and all it has to offer visually oriented students.

Given our individual strengths, weaknesses, and experiences, we all effectively gain, understand, and convey knowledge in ways that are different. The Internet, in an unprecedented fashion, provides us all with the means to dive into the ongoing stream of knowledge and participate in its growth by utilizing and showcasing our own individual assets. As we all begin and continue to expose one another's perspectives, information, and ideas via the Internet, the stream of knowledge is emptying into an ocean with dynamic swells and depths. As we will see, this new ocean has far-reaching capabilities for creating, growing, transporting, and perpetuating its intricate currents of human knowledge.

The act of collecting scientific data has been greatly augmented by Internet use. Taxonomic, phylogenetic, and molecular databases, aimed at organizing and recording the natural world for scientific study, are readily accessible via the Internet. The collection and display of these vast amounts of data are accomplished only through the collaborative efforts of laboratories across the globe. GenBank (www.ncbi.nlm.nih.gov/Genbank/index.html), for example, has become a central depository for DNA sequences. As part of the National Institutes of Health genetic sequence database,

> GenBank is an annotated collection of all publicly available DNA sequences. A new release is made every two months. GenBank is part of the International Nucleotide Sequence Database Collaboration, which is comprised of the DNA DataBank of Japan (DDBJ), the European Molecular Biology Laboratory (EMBL), and GenBank at the National Center for Biotechnology Information. These three organizations exchange data on a daily basis. Each GenBank entry includes a concise description of the sequence, the scientific name and taxonomy of the source organism, and a table of features that identifies coding regions and other sites of biological significance, such as transcription units, sites of mutations or modifications, and repeats. Protein translations for coding regions are included in the feature table. Bibliographic references are included along with a link to the Medline unique identifier for all published sequences. (www.ncbi.nlm.nih.gov/Genbank/GenbankOverview.html, p. 1 [accessed 01/20/03])

Other examples of massive collaborative projects include the Tree of Life Web Project (http://tolweb.org), an Internet project produced by biologists from all over the world aimed at providing the latest information about the characteristics, phylogeny, and biodiversity of organisms on Earth. Similarly, the Centres of Plant Diversity (www.nmnh.si.edu/botany/projects/cpd/index.htm) is an international collaborative project, managed by the Department of Systematic Biology–Botany of the Smithsonian Institution, containing accounts of nearly 250 major localities for the conservation of plant diversity worldwide. Comprehensive descriptions of the geography of localities is provided, along with vegetation, flora, threats, and conservation data.

Intriguing for students of all ages, ARKive (www.arkive.org) is a centralized digital library of films, photographs, and associated recordings of the Earth's endangered species. This grandest of virtual conservation efforts, a not-for-profit initiative of the Wildscreen Trust (www.wildscreen.org.uk), is committed to finding, sorting, cataloging, and copying the key records of species and building them into a comprehensive and enduring audiovisual record accessible via the Internet.

The amount of stored information easily accessible from the types of collaborative databases mentioned here is beyond huge; their maintenance and growth could never be sustained with conventional publication designs. The Internet has clearly made the effective development and use of several immense informational databases such as these possible.

But now that they have been assembled and continue to grow, what are we to do with these gigantic informational databases? Well, as I see it, the aim should be not just to identify and store the information but also to articulate the many different scientific questions that we can ask given such data. These databases contain the raw material for formulating quantitative and qualitative comparative studies of biological systems, helping us elucidate contingent relationships among living things, be they physiological, ecological, geographical, biochemical, or phylogenetic.

Collaborative projects that focus on various microscopic and macroscopic aspects of a study group can lead to more holistic understandings and explanations of nature. The Internet alone has provided researchers with both a rapid and an effective means of conducting such investigations. Laboratories from across the globe and even to and from the International Space Station can stay in contact more or less constantly, exchanging information within intervals of days, hours, or even minutes. Reports and analyses of findings may be discussed and advanced in real time in addition to traditional and formal presentations at annual symposia or in published correspondence.

Annual gatherings at scientific symposia continue to provide researchers with the opportunity to present freshly analyzed data, their latest discoveries, and their new theories. In addition to recent findings, advancements in experimental techniques are also discussed and reviewed at these gatherings, which are developed specifically for information exchange. Since my early days as a graduate student, I have come to view these events as critical to promoting the advancement of scientific research and ideas. However, the scientific gains made possible through such geographically (and sometimes intellectually) localized symposia do have limitations. Often, scientists (and especially students) attend only one or two gatherings a year because of restricted time or travel expenses. Meetings held at different international cities can also make attendance difficult because of limited travel means or language barriers.

Encouragingly, however, voice and video conferencing over IP (Internet Protocol) networks is radically increasing the number of attendees at such gatherings and helping to overcome geographic, language, or financial barriers. The communication platforms offered by some sites, such as Blackboard.com, Centra.com, and Interwise.com, are truly astounding devices for fostering human connections. They provide virtual meeting rooms with all the tools necessary (including multilingual functionalities) to conduct conferences, seminars, broadcasts, symposia, lectures, and tutoring sessions. Through such advancements (which up to now have benefited the business world more so than academia), the Internet can effectively make the participation at scientific symposia a realistic prospect for all students, professional researchers, and amateur scientists who share a vision of the future opportunities and insights that may be made possible by the continued development of science and technology.

In addition to enhancing the capabilities for collaborations at scientific symposia, the Internet is also providing a means for making the published results of scientific research more accessible and useful to scientists and students. Among the Internet content developers leading this venture is an international assembly of scientists who make up the nonprofit Public Library of Science (www.publiclibraryofscience.org). These developers aim to publish scientific online journals that "retain all of the important features of scientific journals, including rigorous peer-review and high editorial standards, but will use a new business model in which the costs of these services are recovered by modest fees on each published paper" (Gordon and Betty Moore Foundation, 2002, p. 1). As leaders of the Public Library of Science initiative explain, open access to scientific publications will accomplish the following:

1. Greatly expand access to scientific knowledge by giving any scientist, physician, student—or anyone with access to the Internet, anywhere in the world—unlimited access to the latest scientific discoveries
2. Facilitate research, informed medical practice, and education by making it possible to freely search the full text of every published article to locate specific ideas, methods, experimental results, and observations

3. Enable scientists, librarians, publishers, and entrepreneurs to develop innovative new ways to access and use the information in this immensely rich but highly fragmented resource. (www.plos.org/about/index.html, p. 1 [accessed 06/30/03])

A supporter of this venture, Thomas R. Cech, president of the Howard Hughes Medical Institute suggests that

> Web publications that are instantly available for free and are readily searchable and downloadable . . . are clearly the wave of the future in terms of our investigators disseminating their research discoveries and learning from the findings of others. In addition, we have a strong commitment to international science and the current subscription systems puts many journals out of the reach of our colleagues in poorer countries. (Gordon and Betty Moore Foundation, 2002, p. 1)

It is my view that as more and more of us depend on web-based publications or symposia for collaborative support, data collection, presentation, and analysis, the overall content and scope of research projects are logically changing in tandem. No longer is the work of students confined only by the materials of their laboratory or the knowledge of their primary research adviser. The Internet is constantly increasing the facilitation and tempo of learning. As such, students can concentrate more on applying rigorous methodologies to answer questions that they themselves find to be important and interesting rather than those that are strictly related to a particular study group, immediate research facility, or faculty specialty. Moreover, students can build a team, or an "active network," of other researchers through Internet associations and effectively work to achieve holistic solutions and explanations.

In building a network of researchers, the significance and implications of interdisciplinary studies becomes obvious. While contributing their own knowledge and skills to an investigation, students can explore their interests further by learning the techniques and perspectives of different approaches and disciplines. Already, colleges and universities understand the educational benefits of interdisciplinary studies, establishing dual degree programs and interdisciplinary majors. Many students today are seeking out such programs in an effort to increase their job prospects. Others are seeking rigorous academic challenges, while others simply cannot make up their minds as to which field to focus on.

Regardless of student rationales, the movement toward interdisciplinary studies is real and has great momentum.

Analyzing problems in isolation cannot lead to comprehensive solutions. Combined disciplines are not only helping provide holistic solutions to large-scale problems that may affect diverse aspects of science, technology, economics, health, transportation, or environmental stability but are also posing new questions, perspectives, and possibilities for the advancement of human knowledge. Combined disciplines now exist within and across every traditional field of study, including arts, humanities, social sciences, engineering, mathematics, and life and physical sciences. Simple browsing through college program catalogs will demonstrate this trend. Materials science and engineering, for example, is a combined scientific discipline that investigates the connections between the structure and properties of various materials, be they metals, ceramics, electronic materials, or biomaterials. The Department of Materials Science and Engineering (MSE) at the Massachusetts Institute of Technology describes the field as follows:

> MSE grew out of the disciplines of metallurgy and ceramics and now includes polymers, semiconductors, magnetic materials, photonic materials, and biological materials. A materials scientist studies how materials react in different conditions (such as temperature and pressure) and understands that all materials can be approached from a common set of principles. Architects, civil engineers, toy companies, NASA, bio-medical research companies, and the data storage industry are all recent users of technologies that developed from research in MSE. (http://tantalum.mit.edu/about/whatis.html, p. 1 [accessed 07/06/03])

Such an interdisciplinary program of study is the embodiment of the future of learning, and I am thrilled about the opportunity I currently have to design an interdisciplinary science course for high school students, the process of which I discuss a bit in chapter 6.

Not only is the Internet making real-time, global, and interdisciplinary collaborations among scientists and students an accessible reality, it is also bringing more and more laypersons in contact with the techniques, overall significance, and implications of worldwide scientific studies. It is becoming possible for us all to come in contact with direct sources of scientific information instead of secondary sources that in many instances are strictly aligned with public opinion. In this event, firsthand scientific information that is within easy reach of the populace is having a considerable effect on the development and maintenance of scientific literacy among world populations.

Today, people are able to gain a better understanding and personal interest in the news of the day as it relates to science. I am strongly encouraged by this development because the reality is that as we begin to head into the depths of the information age, scientific literacy among Americans is surprisingly low despite increases in the number of people with college educations. In fact, scientific literacy is not only an important issue when it comes to the general public's use and understanding of science but also one important to the professional domain. As Hazen and Trefil (1991) speak about in *Science Matters*,

> Intense study of a particular field of science does not necessarily make one scientifically literate. Indeed, it has been our experience that working scientists are often illiterate outside their own field of professional experience. For example, we recently asked a group of twenty-four physicians and geologists to explain to us the difference between DNA and RNA—a basic piece of information in the life sciences. We found only three who could do so, and all three of those did research in areas where this knowledge was useful. . . . The fact of the matter is that the education of professional scientists is just as narrowly focused as the education of any other group of professionals, and scientists are just as likely to be ignorant of scientific matters as anyone else. (p. xiii)

Within the professional realm, much of scientific illiteracy is due, I believe, to reductionism and limited collaborations (see chapters 3 and 4). Within the general public, much of this is due to perpetuated misconceptions in scientific education (Driver, Squires, Rushworth, & Wood-Robinson, 1994), current political agendas, and an essential lack of contact between the general public and high-quality, scientific sources of information. Information that is practically unreachable because of physical remoteness or required levels of comprehension cannot be acknowledged or generally appreciated by laypeople. The Internet, however, is virtually putting all of science's advances within the world's reach. People can readily view the latest images from the Hubble telescope, search the endless databases of biodiversity, study changes in atmospheric conditions, learn about trends in global warming, and inquire about latest discoveries in human genomics and pharmaceuticals. Furthermore, as people become familiar with basic scientific techniques and information, they may gain a level of comprehension and appreciation for methodologies, theories, and implications that is applicable to making advancements in the quality of our intellectual, social, economic, and political goals.

With the development of Internet content, increases in the pace and abundance of scientific information exchange is inevitable. Already, we are caught up in the consequences of the information age—we want information, and we want it now. Students in particular seem to expect to spend less time today searching for information. Throughout my mentoring experience, I have repeatedly noticed that if a student cannot find answers quickly, he or she may prematurely resort to personal defeat or blame an instructor, course objectives, or the quality of resources. This attitude toward learning is hasty, unreasonable, and entirely unproductive. But as an educator of science, I must acknowledge that information exchange has definitely become more rapid and instantaneous because of the Internet. Students today are members of the first generation to use the Internet as the primary and most prominent source of information. As such, I aim to help students focus not on single forms and sources of information for fact retrieval or knowledge growth but rather on the interdependencies among different sources. The importance of this shift in our learning focus from "academic content" to the "context" in which it resides and is accessed is the focus of the upcoming chapters.

Whereas before the development of Internet content knowledge was acquired mainly through individual, linear ways of seeking, gathering, and processing information, now knowledge resides within interdependent and collaborative online networks of information developed by teams of researchers and based specifically on computer and Internet technologies. We must now learn to acquire knowledge from the ever rapidly emerging levels of information—from linked hypertext that transcends the significance of single topics, researchers, or study groups.

I always find it wasteful and rather annoying when graduate level students in particular are fixated in a learning style that relies predominantly on following someone else's instructions. That is, the students work in a lab and follow their advisers' instructions to a T—instructions that may fully encompass laboratory protocols and experimental designs as well as a student's entire research project and perspectives. How, I ask, is the student learning in this situation? Unfortunately, there are too many students who, from formulating a hypothesis to performing tests to understanding the overall significance of the work, never truly come to know the implications of their scientific endeavors and cannot function as scientists outside their specifically confined lab environments or away from their mentors. In making this point, I am not arguing against the work of specialists within scientific fields. Rather, I am strongly suggesting that students be able to "think" and "do" on their own. All laboratories and mentors have their own strengths and weakness. Integrating Internet content and online collaborations can help supplement the weaknesses and complement the strengths while continually exposing students to the world of scientific discovery beyond the walls of their laboratories.

The exposure to diverse scientific methods and ways of thinking is important to the intellectual development of students at any age. I have come to experience such benefits of online learning this past year with my ninth-grade science class. The school building that my students and I occupy is on the lower east side of Manhattan and was constructed in 1912. Our school, which is a new addition to the New York City Board of Education and is now in its third year of existence, relocated to this building just last year. It is a beautiful, historic building, but it was originally designed for elementary students. Hence, there is no space designated for science laboratories. The rooms do not contain water, electrical, and gas lines or even adequate space for running high school and college laboratory activities and experiments.

Helpfully, laboratory space is being designed and planned by architects and engineers as I write this. But the end result, whatever that may turn out to be (you never know when school budgets are so volatile), is a long way off. My colleagues and I have been incredibly challenged by these circumstances, as you can imagine. Every day has been a new struggle helping students design experiments using only the sheer basics of supplies. It gives me the greatest delight, however, to note that in addition to enthusiastic students and adventurous faculty, the rooms of the building have Internet access. I cannot begin to adequately describe to you the countless ways in which the Internet has helped the students and me not only close the deepest gaps in supplies, space, time, and resources but also advance our view of curriculum standards. These standards, which I discuss more in chapter 6, stand apart from even the most advanced laboratory technology. For now, I will summarize for you how the Internet influenced our learning with just one example that truly demonstrates its far-reaching capabilities.

Given our lab space situation, a colleague of mine suggested that we contact other learning institutions in the Greater New York area and see if we could involve students in some of the various types of scientific work that is currently being conducted and that the students might find particularly interesting. It was such a great idea, and she had tremendous knowledge of the people and places to start contacting. One contact she made was with Cold Spring Harbor Laboratories in Long Island. Their associated Dolan DNA Learning Center is a marvelous resource for students, offering laboratory workshops, field trips, online bioinformatics tools, and links to worldwide genome projects, research endeavors, journal articles, and the latest findings in DNA science. They also outreach to local schools, running hands-on experiments involving DNA restriction analysis, bacterial transformations, human DNA fingerprinting, and mitochondrial sequencing.

When we investigated the opportunity of having members of Dolan DNA Learning Center work directly with our students, I asked the director, "How can this happen; we don't have the supplies to work with?" He explained, "Our educators bring supplies with them. And further study and analysis can be conducted by the students, with members from our learning center, over the Internet." Not only did we have a solution to our "lack of lab supply" problem, but with this single project we encountered a new way to expand the learning horizons of our students.

5

OUT WITH THE OLD, IN WITH THE NEW

The power of the Internet continues to amaze me every day. Over time, we continue to experience the inventive new impacts that its development is having on our daily lives. Information of all sorts is literally at our fingertips, and its access is becoming more and more of an active rather than a passive process. When we access information online, we can easily select among a variety of modes of presentation and interaction: transferable data, text, images, audio, animation, or a combination of such modes. In addition, as the previous chapter highlighted, we are no longer geographically, socially, or economically distanced from sources of information.

Those of us who cannot travel to visit the Library of Congress in person can browse its contents online. We might fetch a copy of the Declaration of Independence, download an audio file of Martin Luther King Jr.'s "I Have a Dream" speech, read chapters from Charles Darwin's *On the Origin of Species by Natural Selection*, view type specimens from the Royal Botanical Garden, peruse articles from newspapers around the world, see the latest images from the Hubble telescope, follow agendas of the space shuttle missions, track the journeys of migrating birds, listen to a movement from Mozart's Jupiter Symphony, communicate with amateur or professional paleontologists about the latest fossil finds, observe live video camera images of major construction and engineering projects, post our own written testimonials or commentaries, or watch an upcoming movie trailer.

With no other medium can we simultaneously combine, access, and share various modes and forms of information, literature, and entertainment. If this does not blow your mind, what will? Just think about it. Like the cosmos and all its galaxies, the Internet and all its endless information modules is bigger and grander than can be imagined. But unlike the cosmos, the Internet is our own creation. We built it and continue to advance its growth and expansion. What a time it is to be alive.

Now that we can access so much of our world's culture, history, and technology via Internet content, what will the development of such content be like in the future? As the Internet grows beyond its infancy, I find it interesting that much of the online content continues to be comprised largely of "repositioned" information: information that has been transferred from its original, static form of publication, such as books, journals, video, or audio files, to electronic configurations that reside within the World Wide Web.

There seems to be an immediate rush to dump and disperse as much information as possible onto the Internet. In addition to hard-copy formats, published information now also comes in an electronic version. Everything—from classic novels, textbooks, encyclopedias, magazines, transcripts, newspapers, and journal articles to instructional manuals, reference documents, lesson plans, calendars, and maps; from musical recordings and movies to museum halls and libraries—all content and exhibitions are now available (free of charge in many instances) for review, consultation, and analysis on the Internet. But why do we need to simply reposition previously published material? Why two formats, original and electronic, that are in effect identical to one another except for the basic manner in which they are displayed and accessed?

Presently, Internet content development (and much of its use) focuses mainly on the quick and easy process of information storage, retrieval, and dispersal. But, as I see it, this focus has extremely limited potential for promoting the advancement of information growth and exchange. Indeed, the Internet is the largest, fastest, and most widespread medium for information transfer, but its utilization can extend far beyond its characteristic size, speed, and universality. Instead of perceiving the sole or prime function of the Internet as an infinite repository in which to dump all published information, Internet content developers and users are just beginning to recognize the true function of the Internet as a tool for building contextual environments that are essential for knowledge acquisition. The full understanding of this function can come only by leaving behind our conventional experiences and techniques of knowledge acquisition as we tap into the innovated, dynamic, and interconnected network of resources on the Internet.

The reality is that the Internet's vast capabilities for content storage and dissemination reach far beyond the limits of repositioning. As we have seen, the Internet is a medium for communication, collaboration, exploration, interdisciplinary connectivity, and the creation of personalized, information modules. Online content is not limited to a static, isolated format but rather is one that is dynamic and context dependent. However, despite these innovative and unique properties of Internet content, many developers and users continue to apply a conventional, two-dimensional approach to the process of online information gathering.

Numerous websites are designed, viewed, and interpreted as independent sources of information. Moreover, hypertext is underutilized by many people (especially students), and "web browsing" (or "Internet surfing") is perceived as a tedious, unfruitful process, leading users in multiple directions, away from their initial query. Instead of dynamic, context-based content, we tend to use the Internet to seek out an electronic version of static content that resides elsewhere in traditional forms of publication. After all, these latter forms are what we are accustomed to consulting (we have been accessing static media sources since the origins of written language). But in seeking out static forms of information and their renderings on the Internet, we are augmenting the practice of repositioning to an entirely different level that not only includes the repositioning of information itself but of methods of research and presentation as well.

Content repositioning is a futile practice when it comes to improving the access, expansion, and utilization of the Internet as an information resource. The Internet offers an unprecedented level of connectivity—connecting people to each other (one to one, one to many, or many to many) every day. The hyperlinked structure of online information is not conducive to repositioned, two-dimensional text or linear methods of organization and research. Instead, the interconnected networks of hypertext, computers, handheld devices, and software constitute an informational infrastructure that exceeds all networks and forms of information storage that have preceded them.

This new network is beneficial to all aspects of communication, collaboration, and information exchange rather than simply documentation and storage. Novel and dynamic levels of complexity may emerge throughout this new infrastructure, providing a more pluralistic, multivariable approach to understanding observed patterns of information and the true underlying mechanisms that generate them: a task that remains one of the biggest scientific challenges.

When I engage the compound networks of information available on the Internet, I cannot help but think of its potential to help us elucidate true patterns of nature and their underlying causes. For example, the pattern of the evolution of species has long been perceived to be the result of a branching process represented by a "tree of life" (that is, a phylogenetic tree) topology. This tree of life is a two-dimensional, bifurcating pattern of branches with each species branch linked to its closest ancestor at the nodes of the tree (figure 5.1). Each terminal branch or node on the tree is, by definition, a species (Smith, 1994). The pattern of species origination depicted on the tree of life results in two or more branches on a segment of the tree where there was only one branch before. When two new species split from a common ancestral species, each accumulates a unique set of traits or DNA sequences.

The definition of species used here is derived from a particular view of the evolutionary processes of speciation involving reproductive isolation between groups. If groups become reproductively isolated from one another, for instance, by the presence of an insurmountable geographic barrier, their members cannot mate with others outside their respective group. Thus, genes are exchanged only between members within groups, not across groups (figure 5.2). As sufficient genetic differences accumulate between the isolated groups over long periods of time (by agents of evolution such as mutation, natural selection, or genetic drift), the groups may eventually become genetically incompatible and, as such, members from one group would be unable to reproduce and produce viable, fertile offspring with members of the other group. Thus, the two groups would be considered different species. But as some have pointed out, the perception that species have evolved from common ancestors primarily by a simple branching process reflective of reproductive isolating mechanisms may be unrealistic (Doolittle, 2000; Legendre & Makarenkov, 2002; Makarenkov & Legendre, in press).

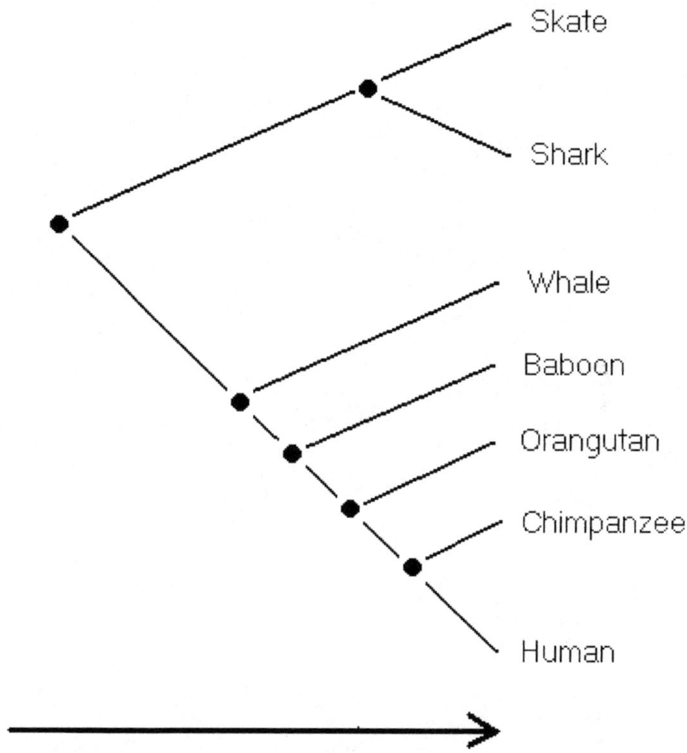

5.1 The phylogenetic relationships of some vertebrates (from Futuyma, 1998, p. 91). The arrow at the right represents time from past to present. The tree-like diagram shows the relative order of the evolution of various taxonomic lineages (branches) from their common ancestors (nodes).

Well-known reticulate evolutionary mechanisms (such as the union of different lineages by hybridization or horizontal gene transfer) that affect the course of evolution profoundly cannot be appropriately depicted by means of bifurcating evolutionary trees (figure 5.3). They instead require an understanding of the fusion of branches (that is, species), culminating in possibly a weblike rather than an openly dichotomous pattern of evolutionary change. As many have argued, reticulate models have been largely neglected mainly because the concept of reticulate evolution has not been supported by analytical tools and software (Alix, Diallo, & Makarenkov, 2003; Makarenkov & Legendre, in press). As Doolittle (2000) suggests in his analysis of the tree of life,

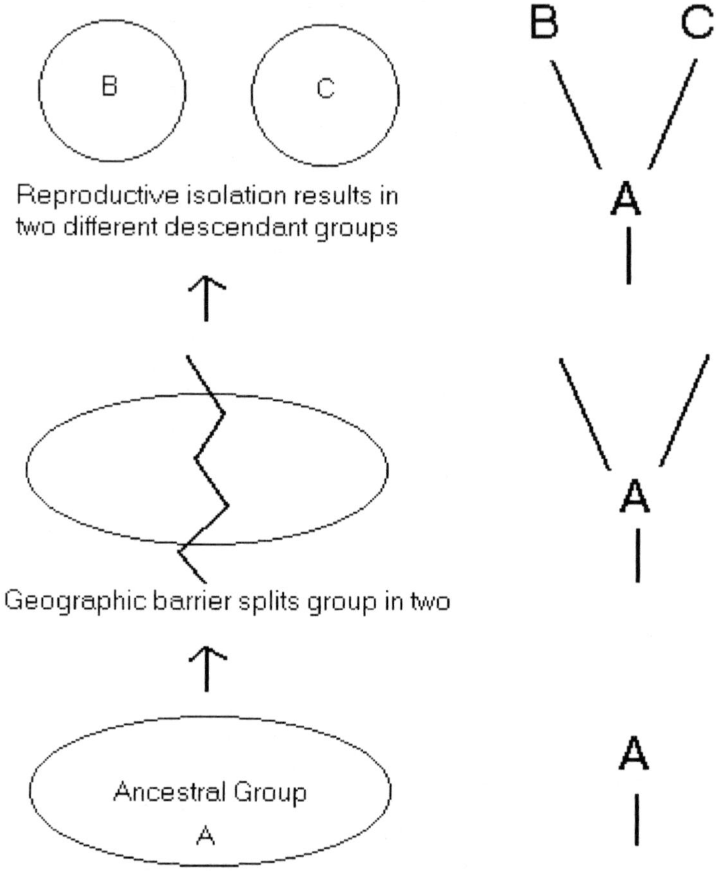

5.2 A simplified illustration of a form of allopatric speciation. An ancestral group (e.g., a species), A, is divided by a geographic barrier. Members on either side of the barrier are reproductively isolated from one another. Over time, they may become genetically incompatible and represent two different species. The left portion of the figure shows how such a speciation process would be depicted on a phylogenetic tree.

based on molecular sequences of representatives from the bacteria, archaea, and eukaryotes, the pattern of evolution is not as linear and treelike as Darwin imagined it. He further points out how some biologists find this notion confusing and discouraging:

> It is as if we have failed at the task that Darwin set for us: delineating the unique structure of the tree of life. But in fact, our science is working just

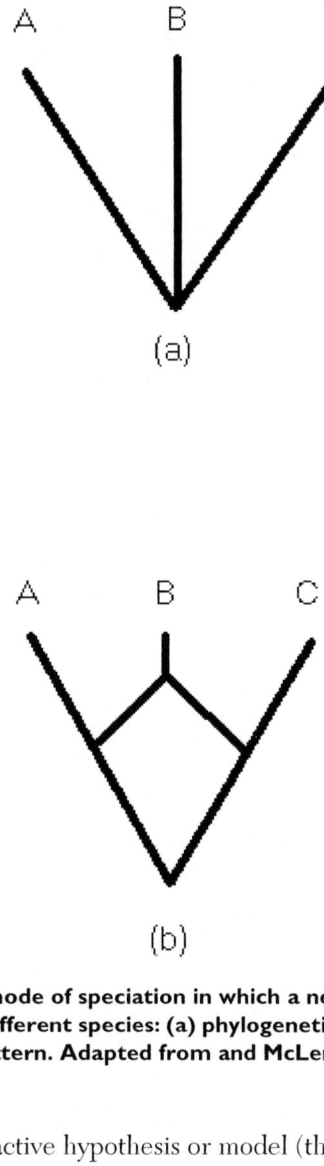

5.3 Hybridization is a mode of speciation in which a new species arises from the cross-breeding of two different species: (a) phylogenetic branching pattern and (b) reticulate branching pattern. Adapted from and McLennan, 1991, p. 106.

> as it should. An attractive hypothesis or model (the single tree) suggested experiments, in this case the collection of gene sequences and their analysis with the methods of molecular phylogeny. The data show the model to be too simple. Now new hypotheses, having final forms we cannot yet guess, are called for. (p. 95)

Following this argument, what quickly comes to my mind is the nonhierarchical structure of the Internet and how decentralized information is within it. It is so unlike any bifurcating model of organization, I often ask myself, Can the multidimensional network of Internet content in any way provide insight to phylogenetic systematists who are always trying to find the true, hierarchical order to the natural world? According to current, mainstream interpretations of Darwin's theory of common descent, the evolutionary tree of life is thought to be openly dichotomous: bifurcating, two dimensional, and hierarchical, not weblike. But I ask, Is our mainstream idea of hierarchical orders of nature real or based simply on our current methods of understanding patterns of speciation and conveying them to one another?

Before the Internet and advancements in software applications, static, written formats were the only means of expressing patterns: We led ourselves to think and perceive two dimensionally because we could straightforwardly design, test, and reconstruct two dimensionally. But with the use of computers and the Internet, multidimensional patterns can be simultaneously conceptualized and conveyed (in the most rudimentary way by the formation of information modules; see chapter 2). My point is that with routine Internet use and development, we may become more comfortable with finding more effective ways to accurately test and describe complex patterns in nature.

As a student of science, I have always been focused on ways in which we can explain both the patterns in nature (those that can be directly observed or inferred) and the processes, or underlying causal mechanisms, responsible for generating such patterns. My own studies that aimed to provide explanations of the origins and evolution of major plant groups involved attempts at describing change in vascular plant form through time as well as explaining the processes that caused the change. In other words, I aimed to describe not only what type of structural change occurred during the evolution of land plants but also how it occurred.

My focus was on the Devonian adaptive radiation of land plants, a time in the Earth's history that is characterized by a dramatic transformation in land plant morphology. During this period (approximately 400 to 350 million years ago), many of the features that we use to interpret relations between modern plant form and function originated, including bipolar and arborescent growth habits, distinct patterns of lateral branch arrangement, leaves, specialized fertile organs (such as cones and sporangiophores), secondary growth (such as wood), and heterosporous life cycles (such as the production of male and female spores). The Devonian fossil record reveals a remarkable increase in morphological disparity and taxonomic diversity that culminated in the origins of all major plant groups, except the flowering plants, by the end of the 50-million-year period (Bateman & DiMichele, 1994). I suggested that such large-scale patterns of evolutionary change that appear to involve successive increases in the variance and complexity of plant form may be characteristic of other major diversification events in the Earth's history, including the expansion of multicellular life (during the Vendian) and the rapid and profound diversification of early metazoans (an evolutionary event commonly referred to as the Cambrian "explosion").

For my study, I used previously published phylogenetic trees depicting relationships among major groups of plants as the basis of several hypotheses used in testing explanations of increases in vascular plant morphological disparity. The tests were conducted by phylogenetic comparative techniques (Brooks & McLennan, 1991; Harvey & Pagel, 1991). These are statistical tests based on correlational evidence. The goal of the techniques is to identify the variable (or variables) responsible for variation in another variable. Comparative tests are based on the null hypothesis that the origin of a trait, characteristic of a species' branch on a phylogenetic tree, is independent of trait origins occurring on any of the other branches. The alternate hypothesis is that the origins of two different traits are correlated throughout a phylogenetic tree. Comparative techniques search for iterative origins of traits along a phylogenetic tree as a means of testing a correlated relationship between the traits. As such, they help elucidate such questions as the following (Brooks & McLennan, 1991):

1. How often did particular traits accumulate in the evolutionary history of a taxonomic group?
2. In what sequence did the traits originate in one lineage compared to another?
3. Did particular traits evolve in a correlated fashion? If so, what is the order of evolution of the two traits?
4. Can one trait be inferred as the independent, causal variable responsible for the evolution of another, dependent trait?

I proposed using various phylogenetic comparative methods to test the hypothesis that a particular key innovation (the independent variable), such as an arborescent growth habit, a distinct pattern of lateral branch arrangement, the development of leaves, specialized fertile organs, or heterosporous life cycles underlies the evolution of high morphological disparity (the dependent variable) in different plant groups. I will not delve into the intricacies of my tests, methods, and findings. (The truth is, I can't bear to rehash it all. At the time, I thought I had the mysteries of vascular plant evolution solved. Heck, I even received an award for best student presentation when I communicated my findings at a national symposium in the summer of 1999.) The reality is that I had just barely scratched the surface of an immense mountain of scientific inquiry. If I were to tackle the same questions again, I would not rely on a two-dimensional, bifurcating organizational scheme of taxonomic groups.

Although any phylogenetic tree is merely a hypothesis of evolutionary relationships, subject to future testing as new data are found and new methods for their reconstruction employed, I am no longer convinced that their bifurcating structures are reliable inferences of historical patterns. Nor am I convinced that they or any other observed or inferred patterns in nature should be the basis of a hypothesis concerning processes of change. Process and pattern, although related and equally important in explaining natural phenomena, are not necessarily causally connected. Which do you infer from which—pattern from process or process from pattern? The same process might produce several different patterns. Likewise, a single pattern might be produced by several different processes. Furthermore, can we make distinctions between types of processes or identify contingencies between processes and their culminated effects? For example, might there be both proximate and ultimate causal mechanisms of measureable change? These ideas, as vague as they may seem, remind me of the genetic effects of pleiotrophy, polygenic traits, and epistasis, which are aspects of gene expression that underscore the complex relationships between patterns and process of phenotypic development. Are similar effects of various subparts perhaps active at other levels of organization?

I believe what is needed to clarify dilemmas of pattern and process is knowledge of the context of situations: identifiable levels of biological organization and how the subparts of these levels relate and interact to produce frequent and predictable behaviors. Some researchers, such as Strogatz (2003) and Wolfram (2002) (see chapter 3), are already beginning to lead the way toward the elucidation of widespread pattern and process relations in nature.

I now see in retrospect that when I was researching the evolution of terrestrial plant groups, my methods of thinking and analysis were overly simplistic. Notions of complexity, contingency, and context in scientific research did not become apparent or relevant to me until I began to engage Internet content and the process of building information modules during my work as an Internet content editor. My perspectives as a student of science changed through time, as they should. Trial, error, and the continual testing of hypotheses exemplify the meaning of science. Needless to say, my dissertation now sits on a very hard-to-reach corner of my bookshelf.

Returning to the notion of repositioning, I believe that the Internet might help us ask complex context-based questions and provide us with a means of representing and explaining multidimensional patterns and processes only if we use its unique capacity for networking information. If we continue to simply use the Internet as a depository for repositioned material, we will miss the advantages of one the greatest technological creations in human history.

The act of Internet repositioning became most apparent to me when I began teaching at a New York City public school. I was so excited when I heard that some of the other instructors were integrating Internet content in their courses. To my surprise, I found that for most of them this simply involved putting grade books, assignment calendars, class websites, and e-mail listservs online for their students. I must admit that I am perceived as "old-fashioned" by many of my "Internet savvy" colleagues in that I have never used the Internet to store data such as grades and assignments. But to me this is in no way effectively integrating the Internet into course agendas.

I also noticed some colleagues eagerly trying to find ways to "fit" Internet content within their course curricula in an effort to help solve problems associated with student achievement of certain state standards. Again, I must admit that I found many of their attempts futile to a large degree because much of the Internet's content cannot "fit" into conventional curricula. As I have briefly demonstrated in earlier chapters (and as I will further explain in chapter 6), the processes and obtainable results involved with learning from the Internet are quite different from those associated with textbooks and other static resources. Difficulties with helping students achieve new state standards involved with cooperative learning, critical thinking, data analysis, research, and digital literacy cannot be solved by adding repositioned material, software, or teaching methods and tools to a traditional classroom curriculum. Solutions to these dilemmas must come from applying a new approach to learning and thinking. By the end of this book, I hope to show that the Internet has the unique potential to make completely new approaches to learning possible for all students.

Before I became actively involved in teaching full time, I was employed by two different for-profit organizations. The first I have already mentioned: bigchalk.com. The second, brainfuse.com, employed me as a director of online academics. In this capacity, I became well acquainted with some of the benefits that can actually be yielded from repositioning. I helped reposition educational content as well as teaching techniques in the forms of lessons guides, standardized test prep guides, and tutoring sessions online.

These tutoring sessions take place in an online virtual classroom. An instructor works with one or more students in this classroom and uses all types of traditional classroom tools that are now made virtual, such as chalkboards, textbooks, notebooks, work sheets, tests, and quizzes. Students and instructors can communicate with each other in real time through written chat messages and audio transmissions (and, possibly by now, streaming video). Tutoring sessions are customized for individual student needs, and teachers and parents are regularly sent reports via e-mail detailing student progress.

The significance of such online content and technique repositioning is clear: The Internet provides the means of bringing students and instructors together who might not otherwise communicate given geographic or economic barriers. Many high-quality online instruction sites exist today: e-textbooks, online universities, web-based courses, online educational forums, and subject-based tutoring centers. For example, the following programs are fully accessible through the Internet, allowing registered users access from any location. These are a mere sampling of the types of educational resources, commercial and nonprofit, that are now available online:

The University of Arizona Biology Project—www.biology.arizona.edu. This site offers supportive materials needed for student-oriented, highly interactive scientific learning on the World Wide Web. The online materials, such as introductory text, models, quizzes, and visualization activities, are used to support lecture, laboratory, and discussion sessions of general biology courses.

MIT Open Courseware—http://ocw.mit.edu/index.html. This site provides free, open access to the publication of course materials at the Massachusetts Institute of Technology (MIT), such as course syllabuses, readings, lecture notes, assignments, exams, and study aides, for educators in the nonprofit sector, students, and individual learners around the world. Accessible information relates to courses in science, engineering, economics, history, art, literature, and much more. The William and Flora Hewlett Foundation, the Andrew W. Mellon Foundation, and MIT jointly fund this grand and ambitious web-based electronic publishing initiative.

University of Phoenix—www.phoenix.edu. The University of Phoenix offers a combination of online and in-person campus classes that fulfill a variety of undergraduate and graduate degree programs in the areas of business, education, health care, management, and technology. Registered students establish online "learning teams" for each course. Students study with their learning team members over the Internet, utilizing e-mail, online libraries, and newsgroups. Beginning and ending classes are held on campus for presentations, final exams, and instructors' summaries.

Minnesota Virtual Academy—www.mnva.org. The Minnesota Virtual Academy (MNVA) currently provides diverse distance learning opportunities for elementary-level students. Like many other virtual schools (Chaika, 1999), MNVA is particularly appealing to home schoolers. The academy is a public school program that delivers its course work by mail and modem. The chosen K–12 curriculum was developed under the direction of Dr. William J. Bennett. In similar fashions, other virtual schools blend innovative online technologies with traditional state curricula.

E-Tutor—www.e-tutor.com. E-Tutor is a K–12 Internet education program for students, parents, and educators, providing effective tools and educational content that can be used to create customized, online learning communities. The site provides interactive lessons designed by teachers from across the United States. These lessons are aligned with state and national goals and standards in curriculum areas such as language arts, mathematics, science, and social sciences.

Brainfuse—www.brainfuse.com. Brainfuse is an online tutoring site that provides instant, unlimited access to expert tutors in math, science, social sciences, and language arts. In an online environment, students and tutors meet one-on-one or in small groups and communicate with each other by drawing on a virtual blackboard, typing instant messages, and even speaking through two-way audio transmissions.

AllLearn—www.alllearn.org. AllLearn is a not-for-profit distance learning partnership among Oxford, Stanford, and Yale universities that provides college-level, online arts and science courses authored by member faculty. In addition, the site offers free study resources comprised of leading academic websites, texts, and lectures to alumni and other adult learners.

Whyville—www.whyville.com. Whyville is an online community dedicated to advancing scientific learning through exploration and communication techniques. This site provides various online discussions, message boards, and learning activities involving the conceptualization and building of inventions that aim to provide student users with multiple paths toward products and solutions. Whyville is a "real" virtual space. Users navigate in 3-D and can even walk through a virtual town. At the site, students are encouraged to conduct experiments related to the real world. They enter their data and observations directly on the site, analyze it with provided tools, and communicate their ideas and results with each other.

All these online instruction sites and many others like them encourage students to read, respond, think, and learn at their own pace and according to their own needs. But their impact on education and student learning can be much greater and more valuable if they strive to move beyond the limits of classroom repositioning. Virtual classrooms, as well as real classrooms, experience the same dilemmas that have long plagued educational systems:

- How do teachers keep students engaged and interested?
- How do they effectively meet the needs of all students while working in collective groups?
- How do they achieve and continue to strive toward new standards?

Whenever I think about these serious challenges facing educators today, I immediately find myself asking, Why confine learning to a single classroom and instructor, be they real or virtual?

Although confined to a single location (website) on the Internet, whyville.com is one online endeavor that appears to work at breaking through the boundaries of repositioning. But through various sites and forms of Internet content (and the ability to construct information modules), we no longer are confined to single classrooms, schools, websites, or any other delimited learning facility. Tutors and teachers can act as cyberspace guides, helping students explore the endless amounts and different kinds of educational resources on the Internet that can provide specialized learning techniques and content to help meet individualized student needs, interests, and challenges. With this book, I aim to show how this might be done and why it should be done as we strive to expand and improve the capacity of human learning.

The multilevel network of information available on the Internet greatly surpasses the limitations of conventional learning resources in terms of their potential to aid and enhance the acquisition of knowledge. Students now have an additional medium (outside the laboratory, library, or classroom environment) to help build and assess their collaborative, cooperative, and constructive learning skills. While becoming accustomed to the rationale and applications of the scientific method, students may experience ample time for discovering facts, exploring information, and taking wrong turns throughout the online information network.

Despite the continual development of educational pedagogies and curriculum standards, no teacher or group of teachers can be all things to all students. Conveying educational expectations, specifying essential academic standards, and building curricula that meet the diverse needs of all students based on their particular cognitive strengths, cultural, and social backgrounds has remained one of the biggest challenges in education. However, the Internet, in conjunction with a classroom instructor's or research mentor's guidance, can facilitate a student's personal, collaborative, and digital learning skills by matching content and learning perspectives and approaches to individual student needs, interests, and pace.

PART 3

LICENSE TO LEARN

You cannot teach a man anything. You can only help him discover it within himself.

—Galileo Galilei, Italian scientist and philosopher (1564–1642)

6

THE NEW STANDARD OF LEARNING

As we are now beginning to understand, Internet content cannot effectively or efficiently be utilized by way of conventional methods of information exchange. The true meaning of Internet content goes far beyond the actions associated with the repositioning of formerly published, printed resources. As I expressed in the previous chapter, the limitations of repositioned resources and methodologies in terms of boosting information exchange and learning potential remain absolutely unchanged in electronic form. As such, many of the dilemmas and challenges that have continually placed educators and researchers at a loss have not been solved simply by integrating repositioned Internet content into formal learning programs. But as I see it, the unique properties of emergence and evolvability of Internet content (see chapters 2 and 3), not manifested in repositioned materials or methods, may help us overcome obstacles that have for so long hindered and limited our capacities for learning.

The skills and information needed to achieve educational success today are mounting. Critical thinking, problem solving, modeling, digital literacy, collaboration, and effective communication skills have steadily been the focus and challenge of our educational institutions' curricula and instructional pedagogies. The truth is, however, that the instructors and conventional resources available at our schools and universities alone cannot provide students with the adequate information they will need to succeed professionally and intellectually in and beyond the information age.

Learning is a lifelong process that merely transcends through what we perceive as the formal education years. And although we may commonly view the past as a simpler and easier time that had "the right stuff" and proper focus on educational priorities, it is hard to imagine that learning institutions were ever really the sole or even prominent provider of student growth and development. Times change, but people's basic needs and desires do not. We are faced today with the same educational struggles of previous generations. For example, we are still trying to do the following:

- Teach all students to read, write, compute, and gain essential comprehensive skills during elementary years
- Help students become active problem solvers
- Motivate students
- Increase student performance on standardized and other achievement tests
- Help students gain technological literacy and skills
- Help students become reflective, lifelong learners
- Prepare students to be respectful and constructive members of society

But one thing that has definitely changed through time is our perception of the world: The world is definitely getting smaller. People are constantly interacting in real time, either in person or over the Internet, exchanging ideas, perspectives, and information. Especially now, in the midst of the information age, when technologies and their means and impact are changing so rapidly, schools and teachers cannot be all things to all students, preparing them adequately for the challenges of a swiftly evolving climate of human perspective, intellect, and design.

With human knowledge and prospect expanding in tandem with that of Internet content, individual teachers cannot be viewed to any extent as adequate "answer dispensers," nor can repositioned electronic information be viewed as "new" or "better" than its static primary sources. It is my sense also that rote memorization techniques or standardized testing will no longer prepare and assess student ability or potential for educational success. Although these aspects of learning have long been debated (Anderson & Sosniak, 1994; Hambelton, 1996; National Commission on Excellence in Education, 1983), the Internet is completely changing the way in which knowledge is acquired and shared. In this event, I believe that the debate pendulum is swinging away from quantitatively to qualitatively measured learning objectives.

Acknowledging this movement, we must change our approach to learning to that which effectively enhances and expands our knowledge base through Internet use. One such approach, problem-based learning, which has already been implemented by many learning institutions with conventional classroom tools, can be greatly enhanced and extended through Internet use by surpassing what were once perceived as insurmountable limitations of classroom supplies, materials, collaboration potential, and time spent learning.

Problem-based learning has been described by Barrows and Tamblyn (1980) as "the learning that results from the process of working toward the understanding or resolution of a problem" (p. 18). These authors further outline the approach as involving the following sequential stages of investigation:

> The problem is encountered first in the learning sequence, before any preparation has occurred.
>
> The problem situation is presented to the student in the same way it would be presented in reality.
>
> The student works with the problem in a manner that permits his ability to reason and apply knowledge to be challenged and evaluated, appropriate to his level of learning.
>
> Needed areas of learning are identified in the process of work with the problem and used as a guide to individualized study.
>
> The skills and knowledge acquired by this study are applied back to the problem to evaluate the effectiveness of learning and to reinforce learning.
>
> The learning that has occurred in work with the problem and in individualized study is summarized and integrated into the student's existing knowledge and skills. (pp. 191–192)

In my view, problem-based learning is aligned in many ways to other learning-style approaches that have been described as "discovery learning" or "a thinking curriculum" (Delisle, 1997; Fennimore & Tinzmann, 1990). Through all these approaches,

> learning becomes an act of discovery as students examine the problem, research its background, [and] analyze possible solutions (Delisle, 1997, p. 9)

and

> students generate strategies for problem definition, information gathering, data analysis and hypothesis building and testing. (Torp & Sage, 1998, p. 23)

These approaches embrace what scientific learning is really all about. As scientists, we ask questions and effectively search for ways to test hypotheses and solve problems. Interestingly, the problem-based learning approach was originally designed for students of medical sciences:

> Traditionally, medical schools taught doctors by requiring them to memorize a great deal of information and then to apply the information in clinical situations. This straightforward approach did not fully prepare doctors for the real world where some patients might not be able to identify their symptoms or others might show multiple symptoms. (Delisle, 1997, p. 2)

Problem-based learning, however, can better prepare any student with the lifelong strategies needed to successfully solve real-life challenges.

The benefits of problem-based learning can be greatly enhanced by Internet use. In building a problem-based learning curriculum, I argue that the problems encountered by students should frequently cut across disciplines, be relevant to real-world dilemmas, and require ample time for decision making, hypothesis testing, data collecting, experimenting, evaluating, comparing, and understanding causal mechanisms of data and events. By using the Internet to plan, gather, share, and evaluate data, students can acquire content and construct their own information modules through their own associative pathways. Sufficient time for research and collaboration time may be provided through instantaneous online exchanges.

In many instances, the Internet nurtures a desire to ponder and explore real-world circumstances within an interactive environment. For example, by taking information from the National Oceanic and Atmospheric Administration website (www.NOAA.gov), students were able to collect and compile data from the buoys off the coast of different continents and islands. During the 2002 academic year, students from Tupelo Middle School in Tupelo, Mississippi, took monthly readings that measured wind speeds, gusts, wind direction, wave height, dew point, air temperature, and water temperature, mirroring the work of many NOAA scientists (CNN.com Special Report, 2002). At the end of the year, students were able to graph the results and prepare a project about their chosen buoy.

Similarly, Rutgers Marine and Coastal Sciences has built a website (www.coolclassroom.org/home.html) in which marine researchers share real data about the state of the ocean over time with teachers and students. Projects include introductory units on the fundamentals of oceanography using real data and outlining specific steps in the student research process, such as forming a hypothesis, predicting outcomes, and writing articles. Students can virtually join scientists in the "COOL-room" as they explore the waters off New Jersey and view the ocean from various vantages points above the Earth and below the water's surface.

In addition, it's interesting to note that the NASA SCIence Files (http://scifiles.larc.nasa.gov) includes online activities, investigations, and tools aimed to foster problem-based learning. This site accompanies a television series produced at the National Aeronautic and Space Administration's Langley Research Center in Hampton, Virginia, and helps students become critical thinkers and problem solvers by confronting them with real-life problems whose answers are never obvious. This series can greatly help teachers meet individual student needs by letting advanced students put learning into action or directly meeting the auditory or tactile learning styles for students with certain learning disabilities (Butler, 2002). Cognitive diversity among students can be tackled by the use of online activities such as those contained within the NASA SCIence Files. As I discuss in the next chapter, maintaining such diversity is a critical asset to classroom learning.

DNA Interactive is another online learning resource offering students fantastic problem-based learning opportunities. Unlike any other visual learning aide, DNA Interactive (www.dnai.org/index.html), maintained by the Dolan DNA Learning Center at the Cold Spring Harbor Laboratory allows students to do the following:

- Explore some of the techniques used for gene manipulation, forensic analysis, and other uses of genetic information

- Solve phylogenetic mysteries
- Envision our medical future and the immense implications of DNA research
- Understand the history of DNA science
- Come to know the viewpoints of Noble laureates and their contributions to our understanding of DNA

Like other scientific websites with comparable amounts of data and investigative reports, DNA Interactive is constantly being updated with the latest information and news concerning genetic research and discoveries. Through streaming animations, model manipulations, and detailed text, this site surpasses the presentational and contextual limitations of any textbook or video, the content of which becomes dated on publication or production.

As incorporated in the websites just mentioned, problem-based learning emphasizes the creation of many opportunities for students to interact not only with Internet content but also with each other. It promotes collaborative learning such that students working in a group might work on different aspects of a problem, create their own information modules, and seek important, comprehensive, and relevant information from each other that is needed to solve the problem in its entirety. When students work together on a problem, a teacher will undoubtedly note that students from different backgrounds will see different aspects and meanings of a problem. I suggest that, through Internet access, the contextual basis of a problem can be expanded far beyond the perspectives of students in a classroom or even an entire school.

With the Internet, students from the same class or study group can collaborate on a given problem, and information exchange can occur beyond classroom barriers. Through its unmatched potential for global collaborations and communications, the Internet is a learning environment that promotes the exchange of diversity and multiple perspectives. The results of such an extended group effort may greatly enhance student understanding of material:

> When students can relate school learning to important real-life issues, they are more likely to seek and value the perspectives of others—peers, teachers, parents, community members, and experts. In so doing, they

> develop interpersonal competencies for creating and participating in dialogue with individuals who have different perspectives and backgrounds. Thus, they not only connect content to their own backgrounds, but they also learn how different people interpret and organize content based on their different perspectives. As a result, a thinking curriculum builds multicultural understanding while encouraging the philosophical understanding of different kinds of knowledge and the limitations inherent in attending to only one perspective on a subject. Students will thus be better prepared to participate in an increasingly global society. Understanding and valuing multicultural perspectives emerges from dialogue in a classroom that is a community of open and sustained inquiry. (Fennimore & Tinzmann, 1990, p. 4)

My own classroom observations and experiences suggest that young students have a natural curiosity for discovery and problem-based learning. As such, I like to shift the emphasis of curriculum objectives from specific content to those of learning techniques that can be universally applied to different problems so that students can pursue their own interests while meeting essential and rigorous academic standards. The Internet, with its endless content and pathways for building information modules, is the most powerful means to help achieve this goal. I came to realize this by reflecting on my own learning experiences as a student in high school, as an undergraduate, and in graduate school.

As a high school student, my education was based heavily on standardized test preparation and the memorization rather than the application of facts and content. In college, I viewed my professors as all-knowing experts of their fields—they held all the right answers and knew all the important questions to ask. In contrast, as a graduate student, I relished inquiry-based learning, especially research that allowed me to pursue my own interests at my own pace. I enjoyed and found great satisfaction in the experience of finding the answers to complex questions myself through processes of problem solving and decision making. As such, my academic potential, success, and motivation flourished. I directed my own learning and for the first time began using the Internet as a commanding tool to enhance my own ability to research information, ask questions, and seek solutions.

On my own, as a graduate student, I implemented discovery- and problem-based learning techniques. As others have suggested, this self-directed approach to learning can be fostered or conveyed by building Venn diagrams (figures 6.1–6.4), which record one's thinking and help show relationships among facts and concepts visually so that they can be reviewed, organized, and refined (Torp & Sage, 1998)—in essence, this

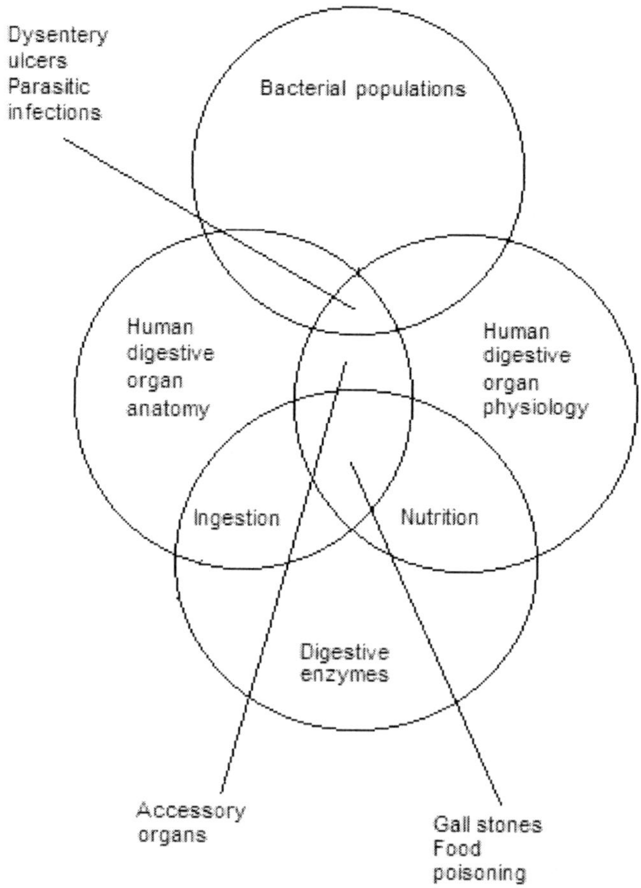

6.1 Venn diagrams allow relationships between multiple sets of topics to be visualized and organized. Topics (represented by circles) are linked by related characteristics or attributes; those that are related are represented by intersecting portions, whereas those that are unrelated are represented by nonintersecting portions. This Venn diagram or "information module" was constructed online by student 1 during a single exploratory search of the human digestive system (see chapter 2).

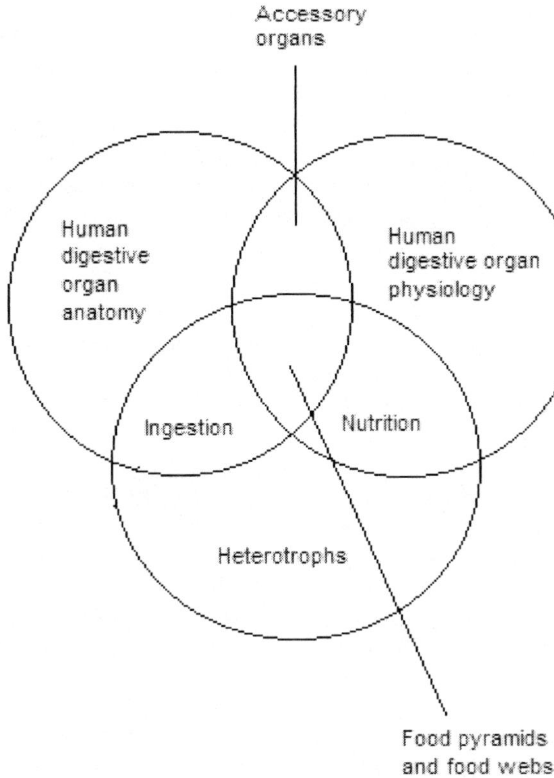

6.2 Venn diagram constructed online by student 2 during a single exploratory search of the human digestive system (see chapter 2).

is the building of information modules (figure 6.5). As I became acquainted with Internet use, I realized that I could "think" and "do" at the same time—I could build multifaceted Venn diagrams, representations of semantic maps if you will, as I actively researched online information. This method of fundamentally combining theory and practice alone can help students engage, master, and perpetuate a universal standard of learning, the sole objective of which is for students to create their own knowledge. They can do this by asking and answering their own questions, ones that they find are relevant to their own understanding of a structure or event.

I cannot emphasize enough the importance of pursing scientific knowledge based on one's own interests and view of the significance of a

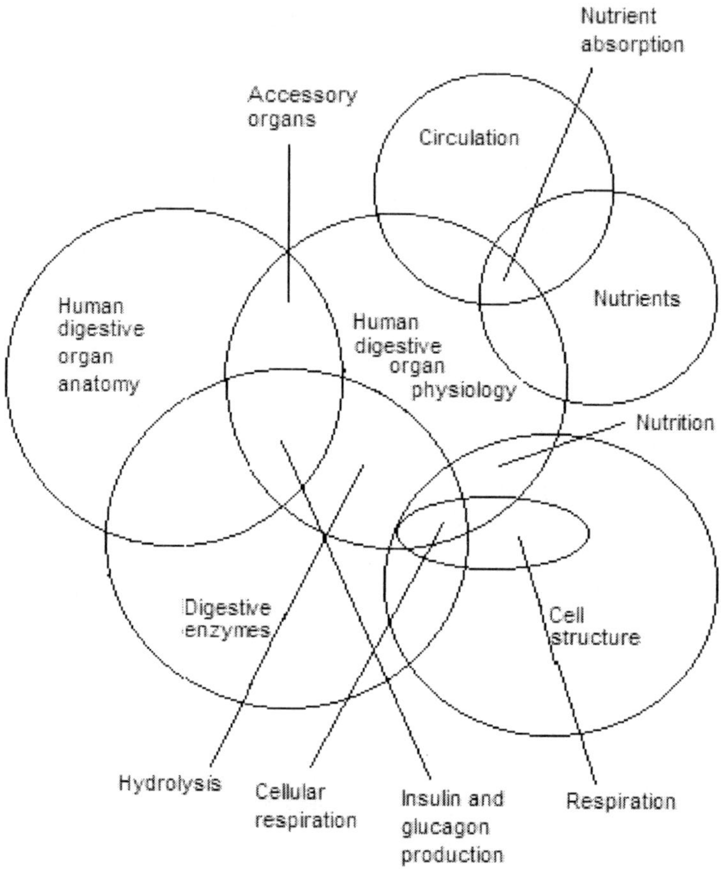

6.3 Venn diagram constructed online by student 3 during a single exploratory search of the human digestive system (see chapter 2).

problem. While working on a committee to build an integrated biology/chemistry/physics course curriculum for high school students this past year, I found myself in a silent, intellectual battle with other scientists and educators over which topics to include and deem essential for student progress in the sciences. The curriculum is designed to provide a survey in the biological and physical sciences through a thematic, integrated approach. Our goal is to help students learn to solve problems scientifically as well as conceptualize and understand the significance of data from different perspectives as well as within multiple contexts and levels of organization. Although all of us consistently agreed on the par-

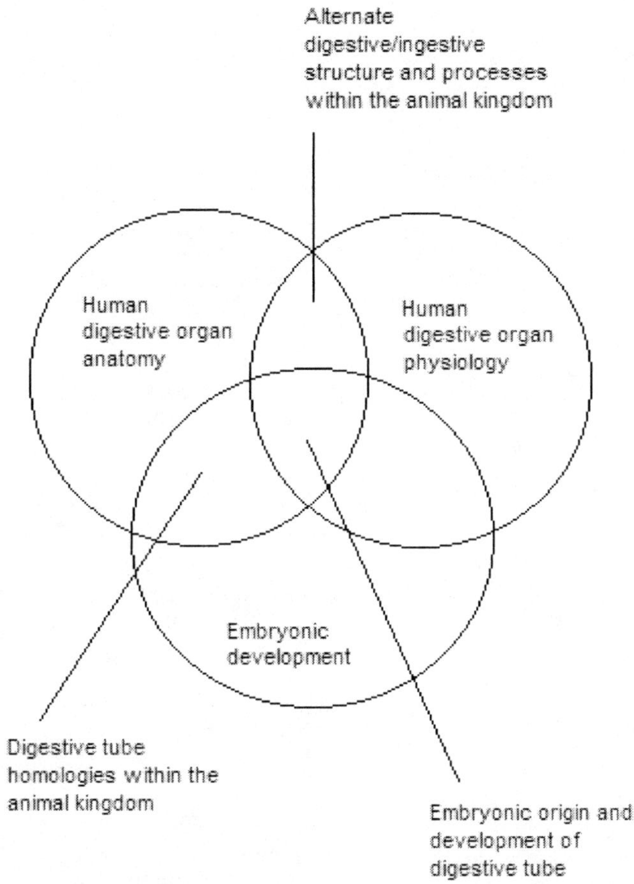

6.4 Venn diagram constructed online by student 4 during a single exploratory search of the human digestive system (see chapter 2).

ticular scientific skills that should be understood and mastered by students, we differed on which topics and themes should be used, how they should be used, and when to introduce and demonstrate these skills.

I think this conflict persists because each of us sees the significance of problems and their solutions differently. But one thing is surely consistent among us: As scientists, we all aim to ask intriguing, provoking, and testable questions. This to me is the the true essence of a successful curriculum or learning standard. Questions drive learning. What I view as an important question may be viewed as completely irrelevant to an-

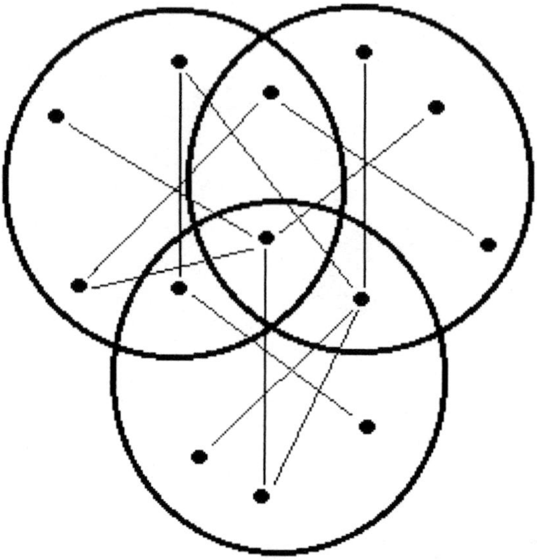

6.5 Underlying an Internet user's information module (depicted here as a Venn diagram) is a complex framework of nodes connected to other nearby nodes. Nodes (shown here as dots) represent individual websites or pages, whereas lines represent associations between nodes constructed by hyperlinks. Nodes and associations comprising the foundation of information modules are different for each user.

other colleague's perspective or research goals. Moreover, the way in which I seek an answer may be different from the methods used by others yet still be scientific. For example, my field of evolution involves the analysis of historical patterns and processes. Thus, much information is inferred from context rather than directly observed. But throughout the experimental journey of finding an answer, common ground may be crossed and significant connections and understanding among disciplines made.

To make successful contributions to scientific research and learning, one must ask one's own questions and find one's own answers. As I have witnessed through my own teaching experiences, by asking and answering questions that students themselves find relevant, they can address many if not all the learning objectives and themes described by formal curricula. For example, during time spent discussing the significance of biomass and energy pyramids, a student asked me, "Dr. Cordi, which came first, autotrophs [organisms that synthesize their own organic food molecules] or heterotrophs [organisms that obtain organic food molecules by consuming other organisms]?" After providing the student with a quick answer based loosely on Stanley Miller's experiments in 1953 and the proposed heterotroph hypothesis, I later reflected on the commanding significance of her inquiry. I thought to myself, "Wow, what an involved question!" If the student, along with teacher guidance, investigated much of the content and methods involved in answering this question, she could fully engage several of the New York State Living Environment curriculum topics, such as the similarity and diversity among living things, cell structure and function, the chemistry of living things, organic evolution, genetics, the continuity of life, and the interdependence of living things. She could also gain essential skills, such as scientifically testing, analyzing, and explaining natural phenomena. Above all, she could engage two of the greatest questions in all of science: what is life, and how did it originate? I have come to recognize that it is not the meticulously defined and described activities, standards, time lines, and objectives of course curricula that guide and cultivate learning but rather the questions that are relevant to a student's own understanding and curiosity.

My own educational experiences described in this chapter reflect a common pattern that has emerged among epistemological studies. K. Patricia Cross (1998), professor of higher education at the University of California, Berkeley, describes the stages of intellectual development as postulated by the epistemologist William Perry (1970):

> The scheme starts at the low end of intellectual development, with students assuming that there is a right answer to every question, and that the answer is known by an authority—namely the professors who are hired to teach them. Students entering college in the early stages of intellectual development have a low tolerance for ambiguity, but they can grant that in some cases we haven't found the answer yet. Their assumption, like ours as a society, is that authorities in research will tell us the answer, and if they don't know it yet, they will eventually discover it. And like students who want quick and unqualified answers, we prefer that the experts make the answers available to us in brief, clear, unambiguous form, such as the three or seven or twelve principles of learning.

> At the mid-level stages of Perry's student development theory, gray areas appear as students begin to discover that authorities often disagree, and that the views of their fellow students often differ from their own. In an effort to resolve these inevitable discrepancies, students adopt an "everyone has a right to their own opinion" stance. (p. 5)
>
> At the highest levels of intellectual development—a stage rarely reached by those who have been studied—there is an affirmation of identity through commitment and self-actualization. . . . But most developmental psychologists are constructivists. They contend that the highest levels of personal development are reached as the person discovers that truth is relative and depends on context. There is not a single right answer, nor is one answer as good as any other. Rather, at the highest levels of development, the individual is able to evaluate truth in terms of the context in which it occurs. (p. 6)

Unclear to me is whether the stages in scientific thinking and learning that students pass through are truly developmental or rather strictly imposed by the teaching methods of institutions and/or by current social convictions (Kuhn, 1970). Regardless of any such debate, lifelong learners know from their own experiences which method of learning works best: Students must construct knowledge and make it their own. They can do this by being encouraged, at any age or academic level, to ask their own questions and search for their own answers. I fully agree with Cross (1998):

> What we know about learning is that people have to find their own answers by working though the pathways to knowledge. Telling people what the "experts" know is not likely to result in the kind of deeper learning that we want to encourage. . . . What we know about student learning is that students who are actively engaged in learning for deeper understanding are likely to learn more than students not so engaged. (p. 6)

The Internet can engage individual students like no other learning tool. Information modules are essentially multidimensional Venn diagrams in motion, helping students gain perspective, develop contexts, and form their own specialized knowledge. But what do students really think about Internet content as a learning aid? How do they really respond to it in the classroom? How do they really use it on their own for study and research? Are they really aware of the virtual Venn diagrams being mapped, and can they construct their own meaning from them? If not, should a method of learning that relies so heavily on Internet content be encouraged? These questions are the focus of the next chapter.

7

FUTURE LEARNING FACILITATORS

My first year of teaching both high school and college students took place at the newly established Bard High School Early College (BHSEC), a joint venture between Bard College and the New York City Board of Education. Students accepted into the school as ninth graders (freshman) finish their high school requirements by the end of their sophomore year and take college-level courses taught by college professors during the remaining two years of study. Ultimately, students have the opportunity to graduate from the school with an associate's degree. BHSEC is dedicated to educating young people in an environment that nurtures academic curiosity and educational excellence, and I am thrilled to have been able to contribute to its mission and goals.

During my first year, I noticed that students displayed rather diverse views about using the Internet as a learning tool. These views seem to mirror those of the general public. Following are some anonymous excerpts of focused free-writing samples that I collected from ninth graders describing their thoughts on the Internet as a learning resource. Students A and B suggest that Internet content is too general and that they would rather be instructed as to what exactly to look for when given an online assignment:

> Student A: "You can get a lot of information from the Internet, but none of it may be useful (too general or not exactly what you were looking for) and you may get incorrect information (there are so many sites on the same things that you don't know what's true)."

> Student B: "I often use search engines such as askjeeves.com, google.com, and yahoo.com but would rather be given websites by my teacher to go to so that I know that I am 'on the right track.'"

In contrast, students C, D, and E express interest in having the freedom to engage the interactive and the diverse aspects of online content:

> Student C: "I usually use the Internet for research, but I also study based on what I find or by chatting online with a friend about subjects. The Internet is an important resource to have. The Internet is interactive, unlike textbooks and workbooks. I also have more reach to other resources online, and am not limited to information in textbooks."

> Student D: "I think that the Internet is a useful resource, but it's hard to find websites about the exact same topics that are taught in class." [Although student D perceives Internet content as unrelated to class content, she also expresses the important point that much of it is complementary to, rather than redundant of, class content.]

> Student E: "I honestly think that the Internet is my most important resource for study and research. It is the easiest way to get the most information. It is also the easiest way to get many diverse opinions about a topic."

Student F writes about how unstructured, confusing, and unreliable she perceives Internet content to be:

> Student F: "I use the Internet to study or do research almost every day. I use it as an extra study aid. I think it's a good and important resource for additional study, but I think it is a horrible resource for a curriculum to be based on. I prefer textbooks, workbooks, and other resources to the Internet because a textbook gives you more structure. You learn the writing style and format, and everything is less scattered. It is much easier to use a textbook! Internet resources can be confusing, as the readings can be for different age-groups and contain unknown vocabulary, and there is false information a lot of the time. Many people put their reports up on a website for a project, but you never know if they are reliable or not. Although the Internet provides different viewpoints, sometimes too many can be incredibly confusing. A textbook might seem less sensationalistic, but it a lot easier to focus on."

Students G and H, however, perceive the Internet as a way to customize information to meet their own learning needs:

> Student G: "I find the Internet to be much more useful than textbooks and workbooks. This is because the Internet sites are more updated. Also, it's so much easier to find exactly what you are looking for over the Internet. Most of the time, in textbooks, the language or wording can be very hard to understand. But on the Internet, you can find a site that's just for you. I use the Internet for every class. I find it extremely helpful. Also, to get certain books to study a topic, you have to buy them or borrow them from the library, but you don't have to do that with the Internet. It is always at home and available. The Internet can also give you sources like newspapers or encyclopedias. It basically has everything you need. It helps me improve my grades by getting extra information within seconds."

> Student H: "The Internet is a truly wonderful thing. I use it almost always to do research, especially if I have to research a large, elaborate topic. Usually, I use the Internet by searching through sites my teachers have recommended. Otherwise, I'll type in the name of a topic or a subtopic into a search engine. I really believe that the Internet is an important research tool. (It's number 1 on my list!) Yes, it's true that textbooks are also important sources of information. However, with a textbook I feel very restricted as opposed to the Internet, which virtually presents an entire world at your fingertips."

Students I and J comment on a popular misconception among both students and educators alike: The Internet is an educational source that chiefly stands apart from other sources in terms of the publication of false or unreliable information. In addition, these students come from the standpoint of "information absorbers" rather than knowledge constructors:

> Student I: "The Internet is a reliable and quick resource, but it has its disadvantages. Many times, the information I get is wrong and unreliable. The more sites you check out, the more differences you will see on specific topics. That's why I believe textbooks should be a main source and the Internet a secondary source. With textbooks, everyone reads and gets the same information and gets the right information."

> Student J: "I used to find the Internet very useful for finding information and still do to a certain extent. But I get very annoyed with the sites that

> pop up when I do a search. Most of the time they aren't useful, and any moron with a computer can write whatever they want about any subject, and you don't know what to believe. I've gotten things wrong on homework assignments because the sites gave me wrong information. I prefer textbooks because you can just look up what you need in the index and get accurate information."

Alternatively, students K, L, and M describe some of the unique communicative aspects of Internet content and the importance of being able to identify reliable online resources and synthesize one's own knowledge on the basis of diverse content:

> Student K: "I use the Internet very often for research and sometimes studying. Most of the time I go to a search engine to search a topic so that I have numerous websites to choose from. I would hate to have to go to the library all the time to do research. I believe that the Internet is a good and useful resource for research because there are many, many good and valid sites. You just have to be wise, use common sense, and look for good reference signs when trying to figure out if a site is a valid one. One class that I study for on the Internet is Chinese. This helps me with my pronunciation because there are many interactive websites for Chinese students."

> Student L: "I prefer to use the Internet because it is quicker and the knowledge is virtually unlimited. I tend to get frustrated switching from book to book to find one thing or study things that are related."

> Student M: "When I study or do research, I am constantly using the Internet. First, I go to a search engine. With the information that I obtain from search engines, I expand on and continue searching from there. I think the Internet is a vital resource because it is accurate and accessible, whereas textbooks may be out of date and are, in my opinion, more confusing and harder to interpret. Also, you can get a vast and diversified amount of information rather than one bias, from one author, from one textbook."

All these student excerpts express very important aspects of Internet content and individual learning styles. We must seriously concern ourselves with the poor quality of content that does exist online and be careful and conscientious when surfing the Internet, identifying clear indicators of web page quality and reliability (Horton, 2000), such as authorship, titles, commercial or nonprofit objectives, home-page associations, copyright dates, and bibliographies. I am still amazed when I hear people suggest that the bulk of Internet content is, in a word, "garbage." The truth is that poor-quality resources are abundant in every form of media—textbooks, journals, newspapers, news reports, and talk shows—many of which often are subject to popular opinion and political agendas. The Internet, by contrast, can provide instant access to unaltered, primary sources of information.

Cross-checking facts, consulting multiple sources, and working as active rather than passive participants in the research process are acts that students cannot become acquainted with too early in their careers. What better, more efficient tool for fact finding and cross-checking than the Internet? A single book can offer only a limited series of outlooks or facts. The Internet, by contrast, contains an endless assemblage of primary and secondary sources for students to browse and assimilate while studying.

Clearly, the publication and broadcasting of misleading information is not a new phenomenon manifested from Internet technology and content development. We must simply recognize and look for the best ways to deal with the amazing fact that all the world is loading information onto the Internet; the bad will undeniably come with the good.

Among students today, there are diverse attitudes concerning Internet use as a study and research tool. Does this mean that using the Internet as a prominent means of discovery- and problem-based learning is futile? If a significant amount of students dislike the Internet and its structure, content, and method of use, how can they effectively learn from it? Much of the variability in attitude among students today may be due to their view of learning in general, its objectives, and what is expected of them as students. From my recent experiences with both college and high school students, I have noticed that many of them enter a class and throughout an entire semester display a manner that seems to constantly beam from them like a loud, bright, blinking marquee: "Tell me what I need to know, and I'll know it. If you don't tell me I need to know it, I won't bother learning it." As educators and scientists, without changing our own view of learning and its methods, objectives, and expectations or demanding that students take initiative and authority in their own educational endeavors, we cannot expect students to change their views and attitudes. From my time spent with students, I have gotten the sense that many are eagerly looking for mentors—professionals to model after and learn from. Catering to the "just tell me what I need to know" mode, even though it does provide a means of working through massive amounts of content throughout a semester, will not help students become lifelong learners. However, helping all students become active participants in and authorities of their own learning will certainly not happen overnight. And the ways in which we might "uncondition" students from the "just tell me" mode are undoubtedly numerous and debatable. But in the meantime, the Internet is here, and its ever-growing content can meet the needs of all students and accommodate diverse learning levels, styles, abilities, and outlooks. We can encourage rather than force students to begin building their own online information modules as they relate to every aspect of course objectives.

As I described briefly in chapter 6, my experiences as a graduate student led me to find that being able to monitor and regulate my own learning helped me view learning as an act that entails all that is meaningful and makes sense. By building information modules, students first set the foundations on their own experiences and knowledge. They learn more by applying what they already know to new situations and content. As associative pathways are explored, the new knowledge that students acquire is meaningful, relevant, and applied.

Application is the most flexible part of my teaching philosophy because it allows the teacher and student to be creative in linking the course content to the student's current experience, awareness, and sense of future. Relevance, on the other hand, allows students to enrich learning and create meaning. The capacities of both application and relevance occur only when students know what to do with content beyond assessment and classroom walls. As discussed in chapters 4 and 5, the Internet can help take students beyond the walls of a classroom, the borders of a state, or even those of a country. Moreover, it can help connect students with other students from different fields and backgrounds to share ideas, collect data, and formulate hypotheses and experiments.

I find that my students in New York City come to the classroom with very different interests, backgrounds, and needs. Although students everywhere have their own particular learning requirements, I believe that the racial, social, and economic differences among the students in a city classroom might help emphasize the significance of "difference" for the instructor. I cannot meet all their needs and interests, but I can help them become self-directed learners: to pave their own paths through the jungle of Internet content and find meaning that is relevant and significant to them.

Students must have access to what they need. Although the Internet contains infinite amounts of content that is customized for each student, how can such content be found? Granted, hyperclicking through online content can create individualized information modules (chapter 2), but is this really a practical and efficient method of constructing knowledge, especially for young or inexperienced students who may have particular difficulty sifting through the online garbage to find sites worth keeping? Here is where a teacher, instead of dictating information, can help guide students through their own cyberspace quests and challenges of information module construction.

There are many sites other than traditional search engines that students can use as the starting point for any online journey. In addition to some of the sites mentioned in chapter 5, EdClicks (www.EdClicks.com) is a collection of some of the best learning websites on the Internet. Also provided are examples of new digital tools that are being used to build and display Internet content and special instructions for beginners on using the Internet to explore online information. The goal of the developer of the site, Judy Breck, is "to help lead students on learning adventures into the vast, interconnected matrix within the Internet." The site is about what is happening to "the cognitive content matrix" that exists on the Internet and the "wonderful new ways each of us can acquire and interact with what is known by humankind!" (Breck, 2003b, p. 1).

The Internet Scout Report (2003) is "one of the Internet's longest-running weekly publications, offering a selection of new and newly discovered online resources of interest to researchers, educators, and anyone else with an interest in high-quality online material" (p. 1). Also available is a searchable and browseable collection of over seven years' worth of critical annotations of carefully selected Internet sites, mailing lists, and other online resources.

By interfacing with the wonders of online content, the "role of teacher" that I promote entails helping students take control of their own learning and education. Traditionally, with problem-based learning, the teacher's role is to create problems that need to be solved (Delisle, 1997, p. 15). But I believe that students can take on this role of "problem creator" as well, expressing which questions are specifically important to them. Successful, engaged learners are responsible for their own learning. These students are self-regulated and able to define their own learning goals and evaluate their own achievement.

In order to have engaged learning, tasks need to be challenging, authentic, and multidisciplinary. All that is interesting, challenging, and relevant to all students does not exist within a single school, college, or classroom, but it does on the Internet. The Internet completely frees students from the limitations of the textbook and an institution's instructional materials and curriculum. There are simply no limits of information resources to research a problem on the Internet. As part of the universal standard of learning that I suggested in chapter 6, I also propose that the role of the student is as an explorer, whereas the role of the teacher is as a learning guide, coach, and coevaluator.

While recently taking on the role of teacher, I have repeatedly recalled a conversation that I had back in graduate school with my research adviser. When I was preparing to present my doctoral defense, he told me that when he was awarded his doctorate, he viewed it strictly as a "license to learn." I believe he meant that learning never stops, even when you are granted the highest of academic degrees, and that with the skills and experience you gain throughout the process, you continually strive to expand your knowledge base and contribute to your field of study. I will always carry this perspective with me.

I have come to view teachers merely as students with accumulated years of learning. Do I know more than my students? In some areas yes but certainly not all. More often than not, when a student asks me a question, my answer is "I don't know." I would never claim to know all that my students need or want to know. Nor would I want them to just take my word as the sufficient end-all to an inquiry. We are all learners.

We all ask questions that are uniquely important to each of us and serve to expand the scope of human knowledge. I have found that students come to greatly respect and appreciate the answer "I don't know" when 1) it is a truthful answer and 2) I help them find their own answers.

My own learning philosophy is based on the concept of "know and do." It is important for students to know the content of any discipline, including the structural components and effective ways of acquiring knowledge. However, knowledge is not enough; the essential questions are, What will students do with the knowledge? and How will they apply what they have come to know through problem solving? I believe that the answers to these questions (which I am constantly looking for) will help students develop into lifelong, self-directed learners.

My style, method, and manner of teaching involves my own active participation in the learning process. As a learning facilitator, a teacher approaches problems *with* students—guiding them through the critical thinking process, which may involve the discovery of facts, generation of hypotheses, design of experiments, and explanation of solutions. Students should experience science as a process for extending their own understanding, not as a process that seeks unaltered truth. This means that teachers must take care not to convey the impression that they themselves or the textbooks, the institutions, or the Internet are absolute authorities whose conclusions are always correct.

My moral concern for student learning establishes a bond with my students. My hope is that they understand that I am a colearner and that they do not view me as an "omnipotent dictator of knowledge." I believe that a student's learning does not come strictly from listening to me or studying for tests. And it is not necessarily reflected by his or her final grade; rather, it comes from his or her fostered ability to acquire and apply knowledge in many different ways.

In my years of teaching college students, I have taken particular notice at how good guessers many of them are. One semester, when I was assisting one of my research committee members who taught undergraduate botany, I was astonished to find that many students who could "ace" a multiple-choice section of an exam did absolutely atrociously when asked to provide a written explanation or a drawing. I was a young teaching assistant at the time, and this turned out to be one of the many experiences I would have with the facades of student learning. Being able to guess at that right answer is not being a self-directed learner; it is not being a student who cares about the content or his or her understanding of a problem's solution or a question's answer: "Problem-based learning calls upon critical and creative thinking by suspending the guessing game of 'what's the right answer the teacher wants me to find?'" (Torp & Sage, 1998, p. 22). "Students, over time, take more responsibility for learning as they develop a set of skills and habits of mind for becoming more self-directed"(p. 64). I am a firm believer in group work as a means of helping students relate to content and to each other while working to achieve critical thinking and process learning skills. For example, I recently engaged my class in an activity, suggested by one of my colleagues, that required students in the class to take turns playing the role of "teachers" and "students." Students worked in groups of four: two working as students and the other two as teachers. The Internet was given as the resource that they could consult if needed. Some students asked me, "Is the Internet the only resource that we can use? Can't we use others?" They did not realize that the Internet contained all that they would possibly need to work through the activity.

This activity, which required the "teachers" to explain the structure of DNA and the processes of transcription and translation, allowed everyone in each group to share their knowledge with each other as well as teach each other relevant techniques and rationales for understanding the material. Not only did this activity help nurture cooperative bonding and a sense of teamwork, but the process skills among students were easily cultivated (more easily, it seemed to me, without my "interference" or perceived "knowledge-authoritative" role).

As I watched the students work together, I observed the conventional roles of student and teacher quickly dissolve, culminating in a more efficient and effective learning system. The roles were gone, and each group worked together as a functional learning unit, asking each other questions, each providing a necessary piece of the comprehensive explanation. I ended up evaluating and assessing the student work on the basis of oral presentations of their findings and debates over related bioethical and medical issues—issues that helped convey to students the significance of the knowledge they have gained to their daily lives. Students learn best when they realize the relevance of solving problems, and effectively explaining this relevance to others is the best demonstration of one's knowledge and potential as a problem solver.

With the Internet, students have access to all the information in the world, but, as I found during the previously mentioned group activity, they at first do not know what to do with all that information. As such, the role of teacher, or what I prefer to call "colearner," is to act only as an experienced guide and learning facilitator, not as an information dictator. As guides, colearners must help students identify quality sites, ask relevant questions, construct their own meaning from them, and steer clear of plagiarism.

Plagiarism by students using online content, one of the biggest gripes made by educators against the use of online resources, is not a fair criticism of the online learning process in my opinion. Plagiarism is not a new trend and did not emerge with the development of online content. It is a problem that has existed since the dawn of education. Granted, it is much easier to electronically cut and paste pages of text, but it is also easier for a reviewer to identify such lifted material in student works (Harris, 2002). The challenge is to help students become self-directed learners and to care about the questions being asked. Only then will the frequency of plagiarism decline. As I have been saying, Internet content can help make self-directed learning happen, but only if teachers keep it in the curriculum and use it effectively and appropriately.

I am still so amazed at the strong attitudes that many educators have against online learning. To me, these attitudes may, in part, be based on ignorance of its use and application or on a preconceived notion that any form of technology is "evil" and leads people astray from their values and needs. The development of Internet content leads to quite the contrary in my opinion. It is one of the most amazing advances in human history, one comprised of a "cognitive matrix where what is known self-connects in context and will soon interface with everyone on the planet" (Breck, 2003b, p. 1). As Judy Breck so eloquently suggests, the goal of Internet users may be "to speed the day when the new presence of useful knowledge on the Internet diminishes the woes of education, closes the digital divide, and completes the dawn we now see of the Global Golden Age of Learning" (p. 1).

This golden age might be characterized by learning systems that are bias free. Teachers today certainly convey their own biases to students, whether intentional or not. Although the authors of Internet content may present their own biases in their works, the information modules constructed by and for individual users are not based on any prior interpretation; they are unique to the user/student.

The golden age might also be characterized by the promotion of cognitive diversity among students and colearners, working together to solve problems. In contrast, tracking students according to their aptitude levels and cognitive abilities is a popular and manageable means of instruction among many learning institutions. Whenever confronted with the option of tracking my science classes, I have always asked myself, "Is this tracking system better for the educators or the students?" Although it is much easier for teachers to teach and manage a classroom of students who are "on the same page" (that is, who share the same level and ability), I do not believe that this is the best way of promoting self-directed learning.

The importance of cognitive diversity in a classroom cannot be overstated. In my opinion, it is a classroom's highest asset. Tracking puts students among their own "cognitive kind." On the contrary, I argue that one of the challenges facing education today is to put the advantages of cognitive diversity into practice within a classroom. When you listen to students working together as a functional learning unit, you constantly hear things such as, "See that?" "Yeah, but you could also look at it this way." "I don't understand; explain it to me again." "Can you explain it another way?" "Who knows how to do this?" "We need to remember to . . .," "How can we test that?" and "What should we use to measure those?" These are the questions and inquiries that guide scientific learning. While guided through the stages of investigation, students encounter the fact that all problems posed to science have two sides—conceptual and analytical. And no matter how flat some of us try to make the "problematic pancake," its two sides remain. Thus, tackling a problem from only one side (that is, engaging only a conceptual or an analytical approach) will not lead to a comprehensive, context-based solution.

Do students have different learning strengths and weakness? Of course. Some minds are more conceptually oriented, formulating grand hypotheses and placing facts and information in various contexts. Others are more analytical, focusing on mathematical measurements and modeling techniques and logical solutions. Will the same students always have the same strengths and weaknesses? Possibly. Our brains are all wired differently. But if we work together on a problem, each contributing in one's own unique and significant way, the greatest of scientific discoveries and explanations are to be achieved. And, as I have been expressing throughout this book, no other tool can bring us together and help us accomplish this ultimate goal more successfully than the Internet.

8

NEW HORIZONS

More and more people from around the world are obtaining access to the Internet every day. This rising trend is real, and, as I have been expressing throughout this book, it is beginning to have an extraordinary influence on the way in which we seek, find, and share information. No longer are slow Internet connections or inadequate telephone lines impeding the flow of information exchange. Now, search engines and software designed specifically for people with slow Internet connections are being developed (Hermida, 2003; Noble, 2003), and the benefits of wireless broadband access are growing at an impressive rate (Breck, 2003a; Negroponte, 2002).

Connections are being made between people, their values, their needs, and their points of view. Discoveries, facts, news, and information are being transmitted through a vast and tightly woven web of hypertext, maintained by an interconnected net of computers. Today, information is spread far and fast, and when we contact it, we also contact countless modules of related data. Information in isolation from collaborations and associations is becoming obsolete. In contrast, the power and meaning of context has never been more influential in our quests for knowledge and human progress.

As I have suggested, the growth and interconnections of Internet content is underscoring for us the shortcomings of reductionist thinking in science. Reductionist observations and theories are limiting, if not misleading, to our understanding of the world around us. Advancing our scientific knowledge of nature relies on a common learning process that places facts within multiple, contextual realms of inference. Although many students and educators have long been aware of the significance of context and inference in accomplishing discrete tasks involving computational, reading, or writing comprehension, I believe that now, in the midst of Internet content development, their significance is becoming more obvious to all of us in our everyday practices of gathering information, acquiring knowledge, and evaluating implications.

Context means everything to our understanding of the natural world: from the structure and properties of atoms to the biotic and abiotic interactions of our planet. Now more than ever, during the dawn of the information age, the pursuit of scientific disciplines in isolation of one another quickly leads to comprehensive dead ends. Alternately, interdisciplinary efforts can provide an infinite foundation for knowledge to grow and progress through time. This idea, as straightforward as it may sound, may not be obvious to everyone, especially students who spend their entire research careers focused on a single study group and set of techniques. For me, however, interdisciplinary study was critical in my efforts to conceptualize and engage research projects.

As a student of paleontology, there was not one question within my research agenda that could be answered by a single field or study of thought. The research entailed the evaluation of evidence from several disciplines: sedimentology, stratigraphy, mineralogy, comparative anatomy and morphology, biometry, phylogeny, palynology, and paleoecology. With the Internet, I could rapidly familiarize myself with particular aspects of these fields, their investigative techniques, and the current research being conducted. In addition, I could read the latest abstracts and contact experts with whom to share ideas and experimental results. Without this permanent and reliable conduit of instantaneous information exchange between myself, other students, and experts around the world, I do not know how I would have successfully pursued the diverse empirical and theoretical studies of my dissertation.

Throughout the processes of learning and research, new intellectual horizons appear in front of us every day, taking on the form of new information, contexts, and perspectives. The Internet creates these new horizons for each user during each journey through cyberspace. As part of the standard of learning that I discussed in chapter 6, I believe that while students ask their own questions about the world and their place within it, colearners can help students engage a research process that truly broadens their intellectual horizons. One valuable outcome of such an endeavor, which brings a student in virtual contact with so much of the world, might be his realization of nature and its order as a scientific wonder that cannot be studied fully from an anthropocentric point of view (that is, interpreting behaviors and functions of living things in human terms only or analyzing their effects and implications as they pertain only to human needs and ambitions). This outlook has a very narrow scope in science, ignoring or misrepresenting the existence and effects of abiotic and biotic interactions that have pre-dated those of humans by millions or even billions of years.

Anthropocentrism may also render a perspective among us that humans are a far superior or highly advanced species while disregarding exceedingly sophisticated attributes of other creatures that clearly extend beyond the natural capabilities of human life: the ability of some bacteria to live at temperatures above 100 degrees Celsius or at pressures up to 16,000 times the pressure present at sea level, or the ability of photosynthetic organisms to make their own nutrients, or the ability of some invertebrates to regenerate body parts, or the ability of some trees to live thousands of years.

In addition, Jared Diamond (1993) makes the suggestion, based on the intriguing taxonomic data revealed by phylogenetic and molecular clock studies of various primate groups, that an anthropocentric outlook inadvertently guides our attitudes about our place in the natural world. He asks,

> Do the new findings about our genetic distance from chimps have any broader implications besides technical questions of taxonomic names? Probably the most important implications concern how we think about the place of humans and apes in the universe. Names are not just technical details but express and create attitudes. The new findings do not specify how we should think about humans and apes. But, just as Darwin's *Origin of Species*, they will probably influence how we do think, and it will probably take us many years to readjust our attitudes. (p. 29)

Broadening horizons and moving beyond anthropocentric perspectives may open doors for students into scientific areas of interest and inquiry that are not obviously connected to human life, history, and goals. Such broadened interests and viewpoints are essential for harvesting the benefits of interdisciplinary studies and expanding the realm of human knowledge. But for scientific inquiry to be truly interdisciplinary, its practices must take into account multiple contexts of information, variable perspectives, attitudes, and research techniques.

I sometimes find it a little alarming that in all my semesters teaching introductory biology courses, I have never met more than a handful of students who have expressed sincere interest in the study of nonhuman life. But I realize that these students are young, and maybe we all start out with the desire to learn about nature as it relates strictly to our own species. Although students may have a genuine interest in human biology (or any other field of study that is human centered), comprehensive dead ends are inevitable without making the connections to other aspects of life and the universe. These connections and contexts are necessary for the advancement of our knowledge and our potential as a species.

By broadening our intellectual horizons through the construction of online information modules, we as learners can easily and readily perpetuate our learning beyond the confines of traditional institutions. By confines, I mean not only those that are pedagogical but also those that are delimited simply by the practical means of school scheduling. For example, college courses meet for only two or three semesters a year. And summer sessions can be impractical because of expense or conflicts with internship or job opportunities. As such, what we perceive as formal, academic learning is fragmented and interrupted.

In addition, I have often heard parents of elementary- and secondary-level students express concern over the length of summer vacations. They are concerned that some students who are not kept in continual contact with academic content quickly forget concepts and skills or lose motivation or know-how. I can understand their apprehensions. Indeed, some skills and concepts require constant practice and application to keep them fresh and relevant. Students also can become anxious, jaded, or even bored while away from academic stimulation. Schools do close for periods of time, but the Internet is always on—in homes, libraries, and community centers. And despite many of the misconceptions about its use, the Internet contains massive amounts of content that is interactive, challenging, intellectually stimulating, and educational. In my opinion, there is no better way to initiate, supplement, and guide learning than through responsible and resourceful Internet use. This applies to all learners: students and teachers.

Although students and teachers alike can become frustrated or bored with their work, I believe that regular contact with Internet content and communicative tools can rapidly reverse many of the feelings associated with "academic burnout." For example, throughout my years in academic circles, I have heard comments from educators such as, "I'm so tired of my subject. I've taught it so many times, and it just gets monotonous and boring after a while." I must admit, although I have had sheer moments of complete frustration while managing classroom- or administration-related affairs and policies, I have never been able to relate to the feeling of academic burnout experienced by so many other teachers. I am constantly seeing the most fundamental of facts or processes in a new light, trying to uncover their connections to other structures or processes. Maybe this is because I am a regular visitor of cyberspace.

On the Internet, there are countless contexts and perspectives for investigating the meaning of facts and ideas as well as their relationships to other types of information and scopes of inquiry. The Internet prevents academic boredom by providing an effective means for considering new and dynamic connections between data and concepts and for readily keeping in contact with the latest discoveries and applications of one's field. For example, I consult several online listservs (obtainable through e-mail) that provide me with the latest news, research, and quality Internet sites about a variety of topics, scientific or otherwise. One such listserv is from the Internet Scout Report (2003; see chapter 7). Another is the Eisenhower National Clearinghouse (www.enc.org), a resource for mathematics and science educators. This online service is created for pre-K–12 mathematics and science educators and provides a searchable online database of curriculum resources and professional development materials. ENC also provides links to the latest news headlines and events in math, science, and technology. And every month, ENC Online's *Digital Dozen* highlights twelve exemplary math and science websites with direct links, comprehensive descriptions of the site's content, and suggested age levels for site use.

Without these listservs, I could never keep on top of the enormous amounts of information that are relevant to my own learning and research. Moreover, I actually enjoy sifting through the latest information and developing new ideas for research, classroom discussions, assignments, and learning activities.

Despite the wonders of Internet content and use that I have been describing throughout this book, many students and teachers have mixed feelings about the Internet—both good and bad (see chapter 7). I think that these conflicting contentions may be characteristic of all people who directly witnessed and participated in historical events involving major technological transitions. Remarkably, we belong to an exceptional cohort that is witnessing one of these events.

We are in the middle of a historic limbo—transitioning from static media sources to those that are dynamic, instantaneously accessible, and far-reaching. We are transitioning from reductionist to emergent ways of thinking, experimenting, and communicating. Along with such transitions inevitably comes conflict: conflict with wanting to dive into the future and relish all its new opportunities and with wanting to hold on to what we have come to know, understand, and cherish. As an expression of this conflict that so many of us are experiencing at the dawn of the information age, there is one more student excerpt from the focused free-writing samples that I presented in chapter 7 that I would like to share here:

> Student N: Most of the time, when I'm on the Internet it is for research, either personal or school related. Most of the time I use search engines. I find searching the Internet to be extremely aggravating sometimes. But when you find something, it can be really worthwhile. I think that the Internet can be better than school textbooks because of the great diagrams it has. But there is always something missing from the Internet that a library has. I'm still trying to figure out what that is. I guess its human contact or the feel of a book in your hands.

This student so simply but effectively expresses the feeling of conflict that many of us share. We want to engage the future of learning but still look back to what has served us thus far. As such, many of us might say, "The Internet is great, but I'd rather learn by reading a book or communicate by writing a letter." I would guess that similar conflicts took place during other times of technological transition in human history, such as those that led to the widespread use of electricity, telephones, antibiotics, automobiles, and computers. But one thing is for sure: We eventually make it through these transitions and grapple with the new problems as well as advance from the new solutions that may arise from them. In the end, we see, understand, and engage the innovative possibilities that technological advancements provide. We finally realize that they do not necessarily replace old ways of living, thinking, and learning but instead provide us with further opportunities for perceiving and exploring new potentials of our existence.

There is a jingle for a Philips Electronics television commercial that I hear once and a while: "I've got to admit its getting better. It's getting better all the time." This expresses such hope, vision, and optimism for the future and all that technology can make possible. When I was enthusiastically discussing the use of Internet technology as a learning tool to some colleagues recently, I recall one saying, "The Internet has its pros, but I can explain more to my students with just simple classroom tools. I don't think anything can replace a classroom with a teacher, students, chalk, and chalkboard." My response was, "That's absolutely right. The Internet by no means can replace what you are talking about." The Internet is in no way a substitute for traditional classroom tools and learning methods. It offers something entirely different—so different that it cannot compatibly replace anything or any method that we have employed before. It instead offers us brilliant new horizons for learning and research.

EPILOGUE

As I am writing this book in my Bronx apartment, I am reminded of the time when I first moved to New York City. I was coming to work with Judy Breck and her team of wonderfully talented individuals to help build the world's finest and freely accessible assemblage of Internet content for students (homeworkcentral.com). My enthusiasm and aspirations for learning were never higher. For the first time, Internet content represented an exciting and innovative resource from which I could learn. And years after my earliest journeys through cyberspace, it still remains so.

Living and working in New York City was a lifelong dream come true for me. And unexpectedly, being a part of city life seemed to provide me with great insight as I pondered the significance and applications of online learning and scientific methodologies. In some ways, this city is like a complex, multicellular organism. But instead of having distinct cells, tissues, and organs, the city functions with well-defined neighborhoods, businesses, and transit systems. Amazingly, the day-to-day contact and interactions of these individual parts culminate in an emergent entity that, to many around the world, represents the epithet of American life. Living here makes me recognize and appreciate that I am an intricate part of a much grander whole.

I feel incredibly fortunate to have found my true callings: to be a lifelong learner, to work with and support the youth of our country, to engage the extraordinary wonders of Internet content, and to help contribute to the indeterminate growth of human knowledge. I also find it absolutely thrilling to be living and participating in such a momentous time in our history. The information age brings to us amazing hope and excitement for the future of human exploration and discovery.

I am curious to think about how this time might be denoted and described by our faraway descendants. Maybe it will be looked on as a second Enlightenment: when individual human experiences and interests rather than authority figures set the foundations for our learning and understanding; when information was far-reaching to and accessible by all peoples; when nature was not analyzed through anthropocentric perspectives; when science and technology freely opened doors to new possibilities, explanations, and solutions while conscientiously taking on the inevitable creation of new problems; and when knowledge of the patterns and processes of life and the universe were gained by an interdisciplinary understanding of multifaceted interactions and relationships among the collective parts of complex systems and a recognition of their distinct, emergent properties.

REFERENCES AND ADDITIONAL RESOURCES

REFERENCES

Alesso, P., & Smith, C. F. (2002). *The intelligent wireless Web.* New York: Addison-Wesley.

Alix, B., Banire Diallo, A., & Makarenkov, V. (2003). New algorithm for improving reticulated networks. http://recomb2003.molgen.mpg.de:9090/poster-04-075.

Anderson, L. W., & Sosniak, L. A. (Eds.). (1994). *Bloom's taxonomy: A forty-year perspective.* Chicago: University of Chicago Press.

Barrows, H. S., & Tamblyn, R. M. (1980). *Problem-based learning: An approach to medical education.* New York: Springer.

Bar-Yam, Y. (1997). *Dynamics of complex systems (studies in nonlinearity).* Cambridge, MA: Perseus.

Bar-Yam, Y. (2000). *Concepts in complex systems: Emergence.* New England Complex Systems Institute. http://necsi.org/guide/concepts/emergence.html.

Bass, N. A., & Emmeche, C. (1997). On emergence and explanation. *Intellectica, 25,* 67–85.

Bateman, R. M., & DiMichele, W. A. (1994). Heterospory: The most iterative key innovation in the evolutionary history of the plant kingdom. *Biological Reviews, 69,* 345–417.

Bentley, P. J. (2001). *Digital biology: How nature is transforming our technology and our lives.* New York: Simon & Schuster.

Berners-Lee, T. (2000). *Weaving the Web: The original design and ultimate destiny of the World Wide Web.* New York: HarperBusiness.

Berners-Lee, T., Hendler, J., & Lassila, O. (2002, April). The semantic Web: A new form of Web content that is meaningful to computers will unleash a revolution of new possibilities. *Scientific American Special Online Issue* : The Future of the Web).

Breck, J. (2001). *The wireless age: Its meaning for learning and schools*. Lanham, MD: ScarecrowEducation.

Breck, J. (2002). *How we will learn in the 21st century*. Lanham, MD: ScarecrowEducation.

Breck, J. (2003a). Education unwired: Development gateway, Afghanistan reconstruction. www.developmentgateway.org/node/134111/sdm/docview?docid=543542.

Breck, J. (2003b). 2003 Showcase of useful knowledge. www.edclicks.com.

Breck, J. (2004). *Connectivity, the answer to ending ignorance and separation: Can you hear me yet?* Lanham, MD: ScarecrowEducation.

Brooks, D. R., & McLennan, D. A. (1991). *Phylogeny, ecology, and behavior: A research program in comparative biology*. Chicago: University of Chicago Press.

Butler, T. P. (2002). NASA reaches out to teachers. Eisenhower National Clearinghouse. www.enc.org/focus/content/document.shtm?input=FOC-002872-index.

Chaika, G. (1999). Virtual high schools: The high schools of the future? Education World. www.educationworld.com/a_curr/curr119.shtml.

Clark, A. (1996). Happy couplings: Emergence and explanatory interlock. In Margaret A. Boden (Ed.), *The philosophy of artificial life* (pp. 262–281). Oxford, UK: Oxford University Press.

"Clever" Project Members (2002, April). Hypersearching the Web. *Scientific American Online Issue*: The Future of the Web), 17–23.

CNN.com Special Report. (2002). Back to school: Innovation and excellence. www.cnn.com/SPECIALS/2002/back.to.school/interactive/gallery.10.things/content 3.html.

Cross, K. P. (1998). What do we know about students' learning and how do we know it. American Association for Higher Education's 1998 National Conference on Higher Education. www.aahe.org/nche/cross_lecture.htm.

Delisle, R. (1997). *How to use problem-based learning in the classroom*. Alexandria, VA: Association for Supervision and Curriculum Development.

Diamond, J. (1993). *The third chimpanzee: The evolution and future of the human animal*. New York: HarperPerennial.

Doolittle, W. F. (2000, February). Uprooting the tree of life. *Scientific American*, 90–95.

Driver, R., Squires, A., Rushworth, P., & Wood-Robinson, V. (1994). *Making sense of secondary science: Research into children's ideas*. London: Routledge.

Dyson, G. B. (1997). *Darwin among the machines: The evolution of global intelligence*. Cambridge, MA: Perseus.

Fennimore, T. F., & Tinzmann, M. B. (1990). *The thinking curriculum*. Oak Brook, IL: Regional Educational Laboratory. www.ncrel.org/sdrs/areas/rpl_esys/thinking.htm.

Futuyma, D. J. (1998). *Evolutionary biology* (3rd ed.). Sunderland, MA: Sinauer Associates.

Gerhart, J., & Kirschner, M. (1997). *Cells, embryos and evolution: Toward a cellular and developmental understanding of phenotypic variation and evolutionary adaptability*. Malden, MA: Blackwell Science.

Gordon and Betty Moore Foundation. (2002, December 17). Public Library of Science to launch new free-access biomedical journals with $9 million grant from the Gordon and Betty Moore Foundation. www.moore.org/news_pr_121702plos.asp.

Hambelton, R. K. (1996). Advances in assessment models, methods, and practices. In D. C. Berliner and R. C. Calfee (Eds.), *Handbook of educational psychology* (pp. 899–925). New York: Macmillan.

Harris, R. (2002). Anti-plagiarism strategies for research papers. VirtualSalt. www.virtualsalt.com/antiplag.htm.

Harvey, P. H., & Pagel, M. D. (1991). *The comparative method in evolutionary biology*. New York: Oxford University Press.

Hazen, R. M., & Trefil, J. (1991). *Science matters: Achieving scientific literacy*. New York: Anchor Books.

Hermida, A. (2003). World's poor to get own search engine. BBC News. http://news.bbc.co.uk/2/hi/technology/3065063.stm.

Horton, S. (2000). Web teaching guide: Finding quality Web sites.www.dartmouth.edu/~webteach/articles/sites.html.

Human Genome Project Information. (2003). Human Genome Management Information System at Oak Ridge National Laboratory. www.ornl.gov/TechResources/Human_Genome/home.html.

Internet Scout Report. (2003, July). Department of Computer Sciences, University of Wisconsin–Madison. http://scout.cs.wisc.edu.

Jensen, E. (1998). *Teaching with the brain in mind*. Alexandria, VA: Association of Supervision and Curriculum Development.

Johnson, S. (2001). *Emergence: The connected lives of ants, brains, cities and software*. New York: Scribner.

Kaufman, S. A. (1993). *The origins of order: Self-organization and selection in evolution*. New York: Oxford University Press.

Kuhn, T. S. (1970). *The structure of scientific revolutions*. Chicago: University of Chicago Press.

Legendre, P., & Makarenkov V. (2002). Reconstruction of biogeographic and evolutionary networks using reticulograms. *Systematic Biology*, *51*(2), 199–216.

Levine, M. (2002). *A mind at a time*. New York: Simon & Schuster.

Lewin, L. (2001). *Using the Internet to strengthen curriculum*. Alexandria, VA: Association of Supervision and Curriculum Development.

Lynch, C. (2002, April). Searching the Internet: Combining the skills of the librarian and the computer scientist may help organize the anarchy of the Internet. *Scientific American Online Issue*: The Future of the Web), 7–11.

Makarenkov, V., & Legendre, P. (In press). From a phylogenetic tree to a reticulated network. *Journal of Computational Biology*.

Mayr, E. (2001). *What evolution is*. New York: Basic Books.

Micklos, D. A., Freyer, G. A., & Crotty, D. A. (2003). *DNA science: A first course* (2nd ed.). Cold Spring Harbor, NY: Cold Spring Harbor Laboratory Press.

National Commission on Excellence in Education. (1983). A nation at risk: The imperative for educational reform (a report to the nation and the secretary of education). Washington, DC: U.S. Government Printing Office.

Negroponte, N. (2002). Being wireless. *Wired*, 10. Wired Digital, Inc. www.wired.com/wired/archive/10.10/wireless.html.

Noble, I. (2003, July 30). Cut down Web for digitally deprived. BBC News, Online Science and Technology Staff. http://newsbbc.co.uk/2/hi/technology/3078243.stm.

O'Neil, D. (2002). Language and culture: Analysis of language. http://anthro.palomar.edu/langiage/language_3.htm.

Oltvai, A. N., & Barabasi, A. (2002). Life's complexity pyramid. *Science*, 298, 763–764.

Perry, W. B. (1970). *Forms of intellectual and ethical development in the college years: A scheme*. New York: Holt, Rinehart, and Winston.

Rorty, R. (1991). *Objectivity, relativism, and truth*. Vol. 1 of *Philosophical papers*. Cambridge, UK: Cambridge University Press.

Smith, A. B. (1994). *Systematics and the fossil record: Documenting evolutionary patterns*. Boston: Blackwell Scientific.

Spitzer, M. (1999). *The mind within the net: Models of learning, thinking, and acting*. Cambridge, MA: MIT Press.

Strogatz, S. (2003). *Sync: The emerging science of spontaneous order*. New York: Hyperion.

Torp, L., & Sage, S. (1998). *Problems as possibilities: Problem-based learning for K–12 education*. Alexandria, VA: Association for Supervision and Curriculum Development.

Watson, J. D., & Crick, F. H. (1953). A structure for deoxyribonucleic acid. *Nature*, 171, 737.

Wolfram, S. (2002). *A new king of science*. Champaign, IL: Wolfram Media.

ADDITIONAL RESOURCES

Alliance for Life-Long Learning. (2002). www.alllearn.org.

Brainfuse.com. (2000). A division of Trustforte Educational Services, Inc. www.brainfuse.com.

Department of Informatics, University of Quebec at Montreal, C.P. 8888, Succ. Centre-Ville, Montreal H3C 3P8, Canada.

Etutor.com. (2002). Strategic Studies Corporation. www.etutor.com.

GenBank Database. (2003). Pittsburgh SuperComputing Center, Carnegie Mellon University. www.psc.edu/general/software/packages/genbank/genbank.html.

Public Library of Science. (2002). Dr. Harold E. Varmus, president of the Memorial Sloan Kettering Cancer Center; Dr. Patrick O. Brown of the Howard Hughes Medical Institute and Stanford University; and Dr. Michael B. Eisen of Lawrence Orlando Berkeley National Laboratory and the University of California, Berkeley, leaders. www.publiclibraryofscience.org.

University of Phoenix. (2002). www.phoenix.edu.

Whyville. (2002). Whyville is a service mark of Numedeon, Inc. www.whyville.com.

INDEX

ABOUT THE AUTHOR

Jennifer Cordi has a diverse background in science education and research. She received a bachelor of arts degree from the University of Rochester and a doctor of philosophy degree in biological sciences from the State University of New York at Binghamton. Her dissertation research included anatomical and morphological reconstructions of Middle Devonian fossil plants and phylogenetic comparative analyses related to the patterns and processes of land plant evolution. Dr. Cordi has spent over three years in the Internet education industry. As chief science content editor at bigchalk.com, she organized an online portal for students and professionals to engage the highest-quality scientific websites on the Internet. She also made significant contributions to the development of online math and science courses as well as study aids, lesson plans, test preparation guides, and other educational resources as director of academics at Brainfuse Online Instruction.

Currently, Dr. Cordi teaches high school science and college-level biology courses at Bard High School Early College in Manhattan, a new venture between Bard College and the New York City Board of Education. She has spent several semesters collaborating with other teachers and science professionals from various New York City high schools, Bard College, and Rockefeller University on a Fund for the Improvement of Secondary Education (FIPSE) project, focusing on the creation of an innovative ninth- and tenth-grade science curriculum that implements an interdisciplinary, scientific approach to learning. In addition, Dr. Cordi is actively involved in the design and development of the college-degree science curriculum at Bard High School Early College.

1551